MAT 103 Introduction to College Algebra

Matthew G. Jones and Sharon Lanaghan

December 16, 2022

Contents

Chapter 1

Linear Functions

1.1 Representing Relationships Mathematically

Goals:

- F: Be able to determine inputs or outputs from a function table.

- F: Be able to determine inputs or outputs from a function graph.

Mathematical relationships between two quantities can be represented in multiple ways: as a table, as a graph or as an equation. Each of these representations adds to our understanding of a mathematical situation, and we should be able to approach a given situation by considering each of these representations.

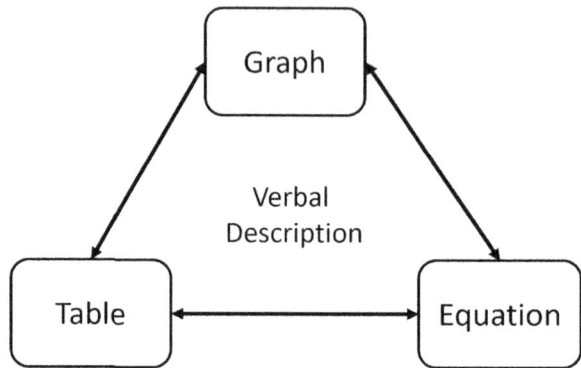

Definition 1. The **domain** of a function is the set of inputs to the function. The **range** of a function is the set of outputs from the function.

Investigation 2. The following table and graph show the relationship between V, the volume of punch in a bowl at a party, and t, the time in hours since the party began.

t (hours)	V (inches^3)
0	4189
0.5	3934
1	3284
1.5	2408
2	1474

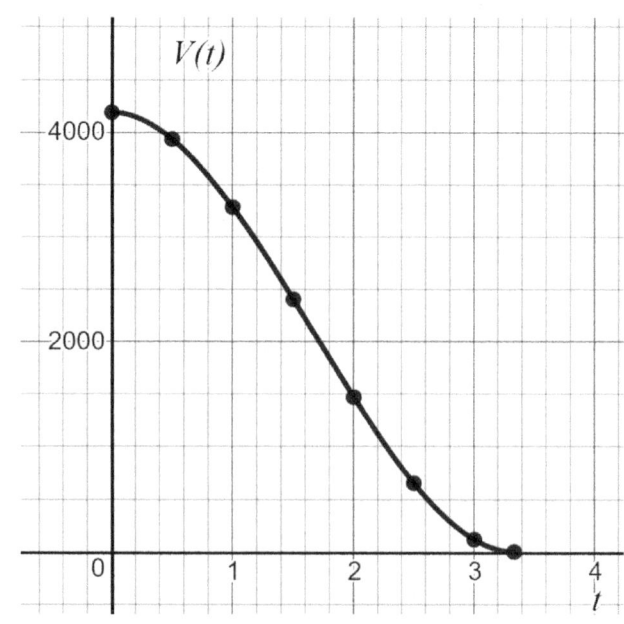

Use the table and graph above for $V(t)$ to answer the following questions:

1. Add two additional points to the table by using information contained in the graph.

2. What is the volume of punch in the bowl 1 hour after the party started?

3. When was the volume of the punch approximately 650 in^3?

4. What was the volume of punch in the bowl when the party started?

5. When will the volume of punch in the bowl be half of the volume when the party started?

6. When will the party run out of punch?

7. In what time interval will the volume of punch in the bowl be between 1474 in^3 and 3284 in^3?

8. When is the volume of punch in the bowl decreasing fastest? When is the volume of the punch in the bowl decreasing slowest?

9. What are the domain and range of V?

10. If the volume of punch in the bowl is given by the equation

$$V(t) = 72\pi t^3 - 360\pi t^2 + \frac{4000}{3}\pi$$

what is the volume of punch in the bowl 1.25 hours after the party started? Add this information to the graph and table above.

Problem 3. The graph below shows the elevation of a hiker, $E(t)$ (in feet), t hours into a 3 hour hike.

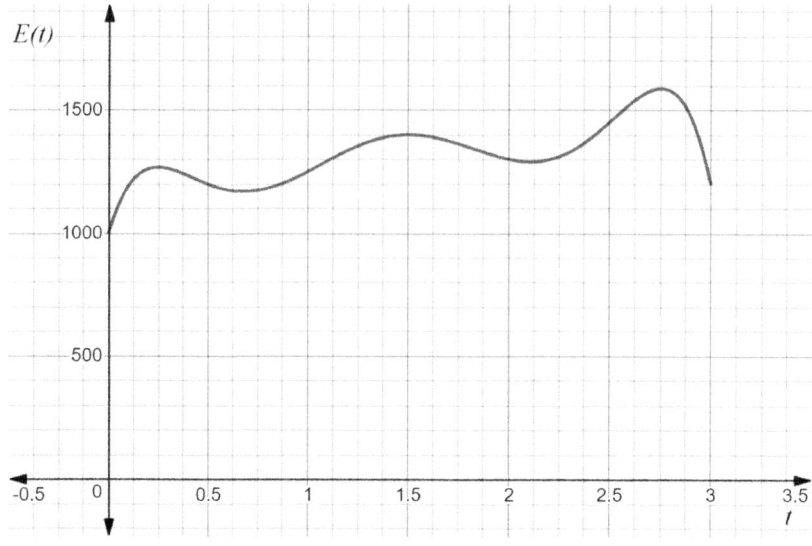

1. What is the elevation of the hiker after 1.5 hours?

2. At what times is the hiker at an elevation of 1200 feet?

3. Solve the equation $E(t) = 1200$.

4. At what times is the hiker at an elevation of at least 1200 feet?

5. Solve the inequality $E(t) \geq 1200$.

6. What is the maximum elevation of the hiker? At what time is he at this elevation?

A function is a specific kind of relation, where each input of the function leads to only one output. Functions are used throughout math and science, and will be a primary focus in this course.

Example 4. Use the graph of g shown below to answer the following questions.

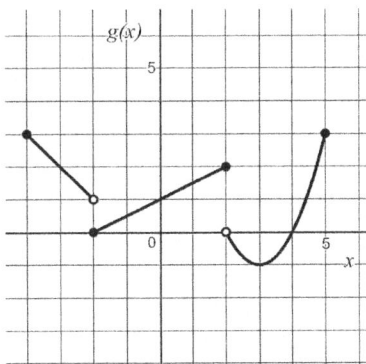

1. Evaluate:
   (a) $g(0)$ (b) $g(2)$ (c) $g(-2)$

2. Solve:
   (a) $g(x) = -1$ (b) $g(x) = 3$ (c) $g(x) = 0$ (d) $g(x) < 0$

Problem 5. The graph of g is shown below.

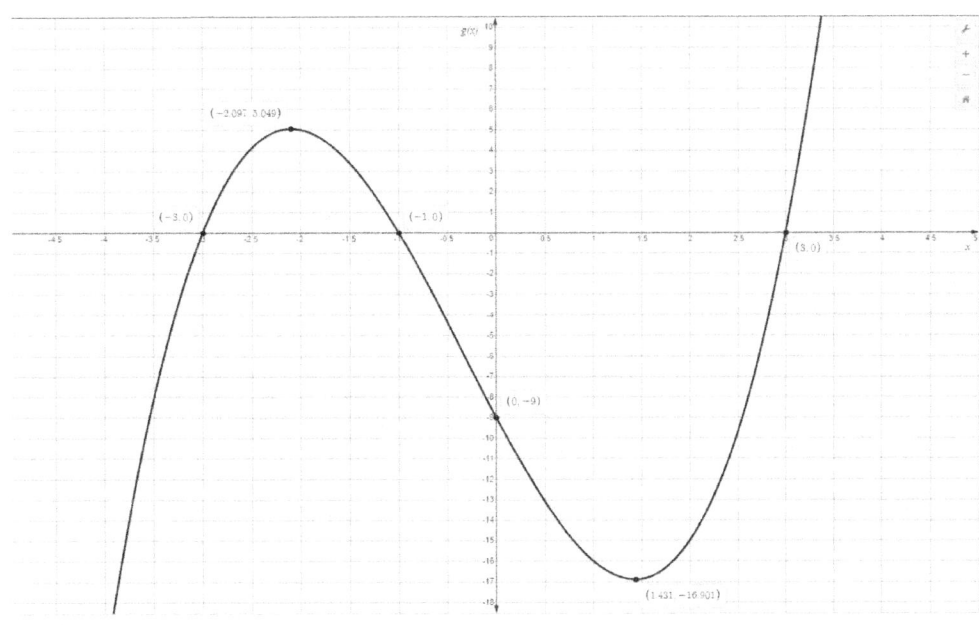

1. Evaluate $g(-3.5)$.   2. Solve $g(x) = 0$.   3. Solve $g(x) > 0$.   4. Solve $g(x) \le 0$.

Problem 6. Use Desmos to graph the function $g(x) = x + \sin(2x)$.

1. Evaluate $g(0)$.
2. Evaluate $g(-\frac{\pi}{3})$.
3. Solve $g(x) = 1.288$.

4. Solve $g(x) = 10$.
5. Solve $g(x) = -1200$.

To determine the output of a function for a given input using an equation, substitute the given input into the equation. Evaluate the resulting expression making sure to follow the **order of operations**:

1. Simplify expressions within grouping symbols (parentheses, roots and fraction bars)

2. Evaluate exponents

3. Perform multiplication and division from left to right

4. Perform addition and subtraction from left to right

Problem 7. Evaluate each function for the indicated value or expression.

1. $k(t) = 3t^3 - 2t^2 + t$, $k(-2)$
2. $q(x) = \frac{x-2}{x^2+2}$, $q(n^2)$

3. $f(n) = 2 \cdot 3^{2n-3}$, $f(\frac{1}{2})$
4. $m(x) = \sqrt{x - 3} + 4$, $m(x^2 + 4)$

Notes

1.1 Exercises

1. The height above ground, h, of a ball t seconds after being dropped off a 60 meter tall building is given by the equation

$$h(t) = -4.9t^2 + 60$$

and shown on the graph below.

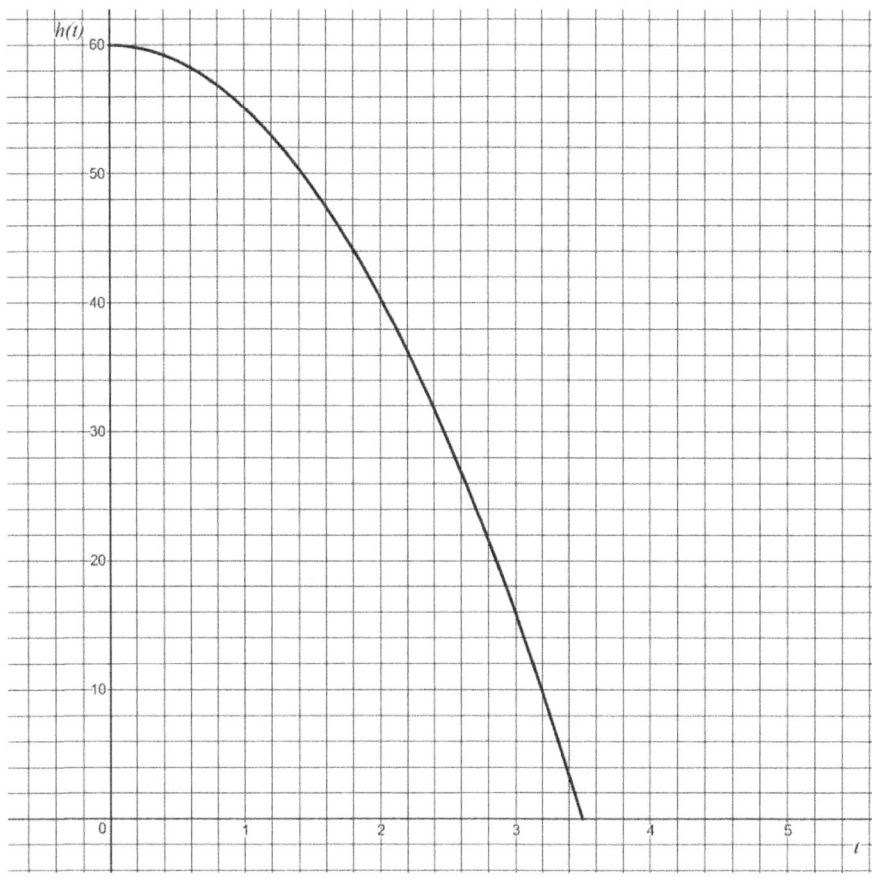

(a) How high is the ball at $t = 2$ seconds?

(b) How long does it take the ball to hit the ground?

(c) When is the ball at a height of 10 meters?

(d) When will the height of the ball be greater than 10 meters?

(e) Using the graph, estimate the height of the ball at $t = 1$ second, then use the equation for $h(t)$ above to find the exact value.

(f) Using the graph, estimate at what time the ball will be a height of 20 meters, then use the formula for h above to find the exact value.

2. Use the graph of $p(n)$ below to answer the following questions.

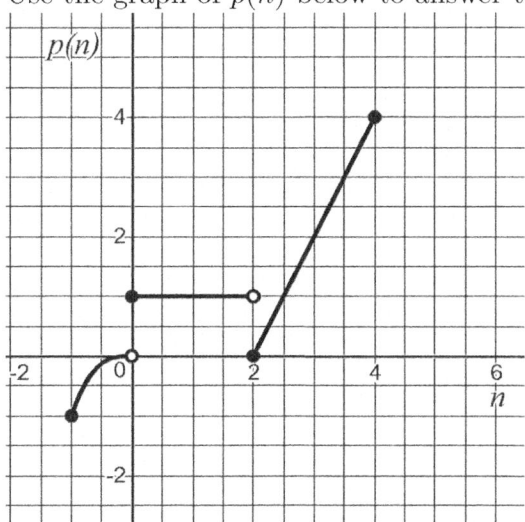

  (a) What is $p(1)$?

  (b) What is $p(0)$?

  (c) Solve $p(n) = -1$.

  (d) Solve $p(n) = 0$.

  (e) For which values of n is $p(n) = 1$?

  (f) Solve $p(n) \geq 2$.

3. Use Desmos to graph the function $f(x) = 4x^3 + 10x^2 - 1$ and answer the following questions.

  (a) Evaluate $f(-1)$.

  (b) Solve $f(x) = -1$.

  (c) Solve $f(x) = 4$.

  (d) Solve $f(x) \geq -1$.

  (e) Solve $f(x) < 4$.

4. Evaluate each function for the indicated value or expression.

  (a) $v(t) = 5t^2 - 3t + 4$, $v(-3)$

  (b) $q(n) = -4 \cdot (-4)^{n^2}$, $q(-3)$

  (c) $b(x) = 5x + 3$, $b(-2x + 4)$

  (d) $z(x) = \frac{x^2+1}{x+1}$, $z(4x)$

1.2   Linear Functions

In this course we will study many types of functions. Linear functions are used to describe patterns where the rate of change is constant, and will be the first type of function that we study.

Goals:

- L: Be able to solve a linear equation.

- L: Be able to model a situation with appropriate linear equation(s) and interpret the solution.

- L: Be able to determine the slope or equation of a linear function given its graph or a table of values.

- F: Be able to determine inputs or outputs from a function table.

- F: Be able to determine inputs or outputs from a function graph.

Definition 8. **Slope** is a measure of the steepness of a line. For a linear function $f(x)$, the slope of a line can be calculated as the ratio

$$\text{slope} = \frac{\text{change in output}}{\text{change in input}} = \frac{f(x_2) - f(x_1)}{x_2 - x_1}$$

Example 9. Find the slope of each linear function below.

1. v as shown in the graph below.

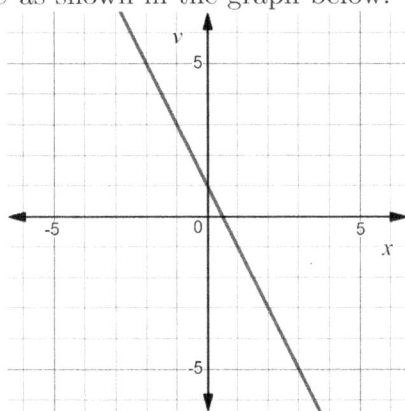

2. b as shown in the graph below.

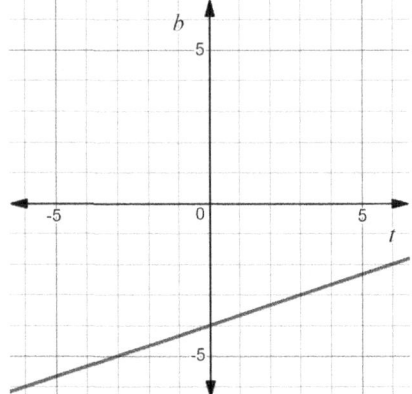

3.

t	-1	0	1	2	3
$g(t)$	2	1.5	1	0.5	0

4.

x	-1	0	1	2	3
$c(x)$	-4	0	4	8	12

Definition 10. Equations of lines are often written in **slope-intercept form**,

$$y = mx + b$$

where m is the slope of the line and b is the vertical intercept (the point where the line crosses the vertical axis). The **point-slope form** of a line is also often useful and is written

$$y - y_1 = m(x - x_1)$$

where m is the slope of the line and (x_1, y_1) is any point on the line.

Investigation 11. Use the graph in Part 1 of Example 9 to answer the following questions.

1. What is the vertical intercept of the graph of v?

2. Write an equation of the graph of v in slope-intercept form.

3. Choose a point on the graph of v. Write an equation of the graph of v in point-slope form.

4. Simplify your answer from part 3. How does this compare to your equation in part 2?

Problem 12. For each function in Example 9, write a linear equation in slope-intercept form and a second linear equation in point-slope form.

Problem 13. For each pair of points below, write an equation of the line in point-slope form. Then write the equation in slope-intercept form and identify the vertical intercept.

1. $(-2, 3)$, $(-1, -1)$ 2. $(-2, -1)$, $(2, 5)$

Investigation 14. For each problem below, use the given representation of the linear function: formula, graph or table, to complete the other two representations.

   1. $f(x) = 2x - 3$

x	$f(x)$
-1	
0	
2	
3	
9	

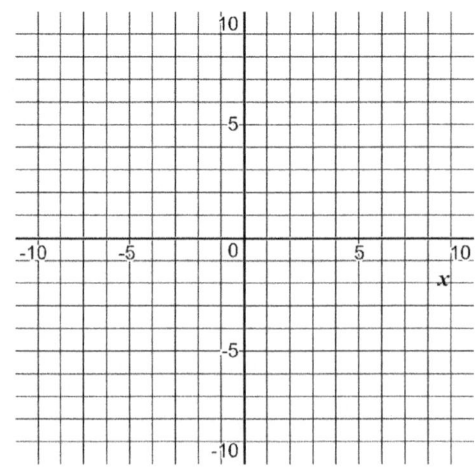

2. $g(x) =$ _____

x	$g(x)$
-2	-5
-1	-2
0	1
1	4
2	7

3. $h(x) =$ _____

x	$h(x)$
-2	
-1	
0	
1	
2	

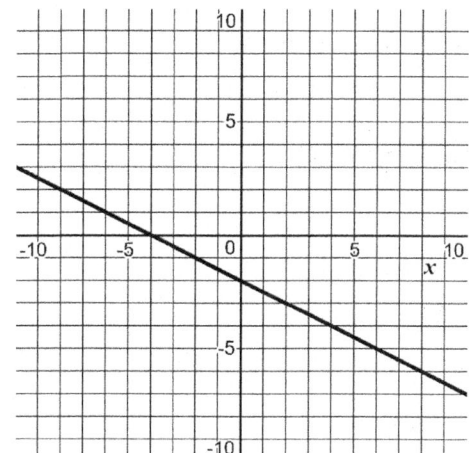

4. Compare the process of moving from table to graph to the process of moving from graph to table: how are these processes alike and how are they different?

5. How do you use the function equation to get one of the other representations (table or graph)?

Problem 15. Let $f(x) = 2x + 3$.

1. Graph the function on the domain $-4 \leq x \leq 4$. Be sure to label the $x-$ and $y-$intercepts and the endpoints of the function on this domain.

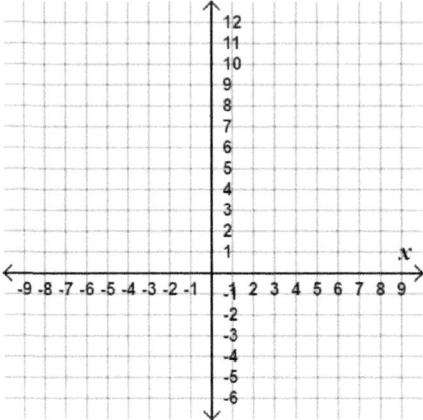

2. Solve $f(x) = -1$ for x.

3. What is $f(\frac{1}{2})$?

Problem 16. Let $f(x) = 6(1 - 3x)$, $g(x) = 2x - 1$ and $h(x) = -x - 2$.

1. Evaluate $h(-3)$.   2. Solve $g(x) = 11$.

3. Solve $f(x) = -2$.   4. Solve $h(x) \geq 4$.

5. Solve $f(x) > 12$.

6. Solve $g(x) = h(x)$.

7. Solve $g(x) \leq h(x)$.

Notes

1.2 Exercises

1. Refer to the graph of $g(x)$ below.

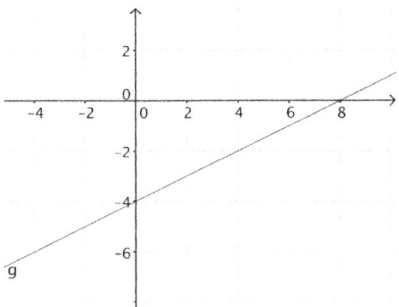

   (a) What is $g(3)$?
   (b) Solve $g(x) = 0$.
   (c) Write a function equation for $g(x)$.
   (d) Use your function equation for $g(x)$ to find $g(3)$. Does this match your answer to part 1a?
   (e) What are some advantages and disadvantages of representing functions as graphs vs representing functions as equations?

2. Let $p(x) = -3x + 12$.

   (a) Graph $p(x)$ on the domain $-3 \le x \le 5$. Label the intercepts.
   (b) Solve $p(x) = 0$ for x using the function equation.
   (c) What is $p(-2)$?

3. For each linear function below, find the slope, and write equations of the line in point-slope and slope-intercept form.

   (a) f as shown in the graph below.

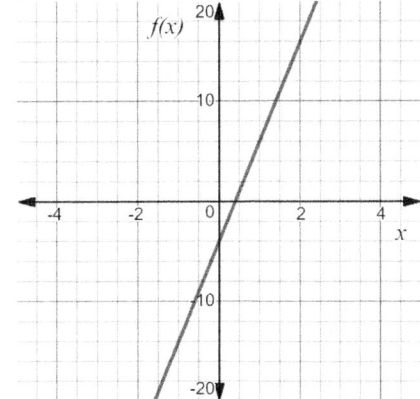

   (b) f as shown in the graph below.

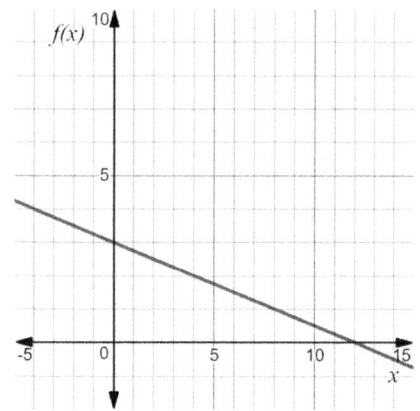

	x	-1	0	1	2	3
(c)	$g(x)$	5	2	-1	-4	-7

	t	-1	0	1	2	3
(d)	$n(t)$	-0.75	-0.5	-0.25	0	0.25

4. Given two points below, write an equation of the line in point-slope form. Then write an equation of the line in slope-intercept form and use your equation to identify the vertical intercept.

   (a) $(-2, -14)$, $(0, -2)$

   (b) $(0, -1)$, $(1, 0)$

   (c) $(-11, 6)$, $(4, -5)$

5. Let $k(x) = 4x$, $j(x) = -3x + 4$, $m(x) = -4(x - 2) - 3$.

   (a) Evaluate $m(-2)$

   (b) Evaluate $m(0)$

   (c) Solve $j(x) = -2$

   (d) Solve $k(x) = m(x)$

   (e) Solve $j(x) \leq -1$

   (f) Solve $m(x) > 0$

   (g) Solve $m(x) < k(x)$

6. The graph below shows the distance above the ground of a rider (rider's height) on a children's Ferris wheel, d (in feet), t seconds after the ride begins.

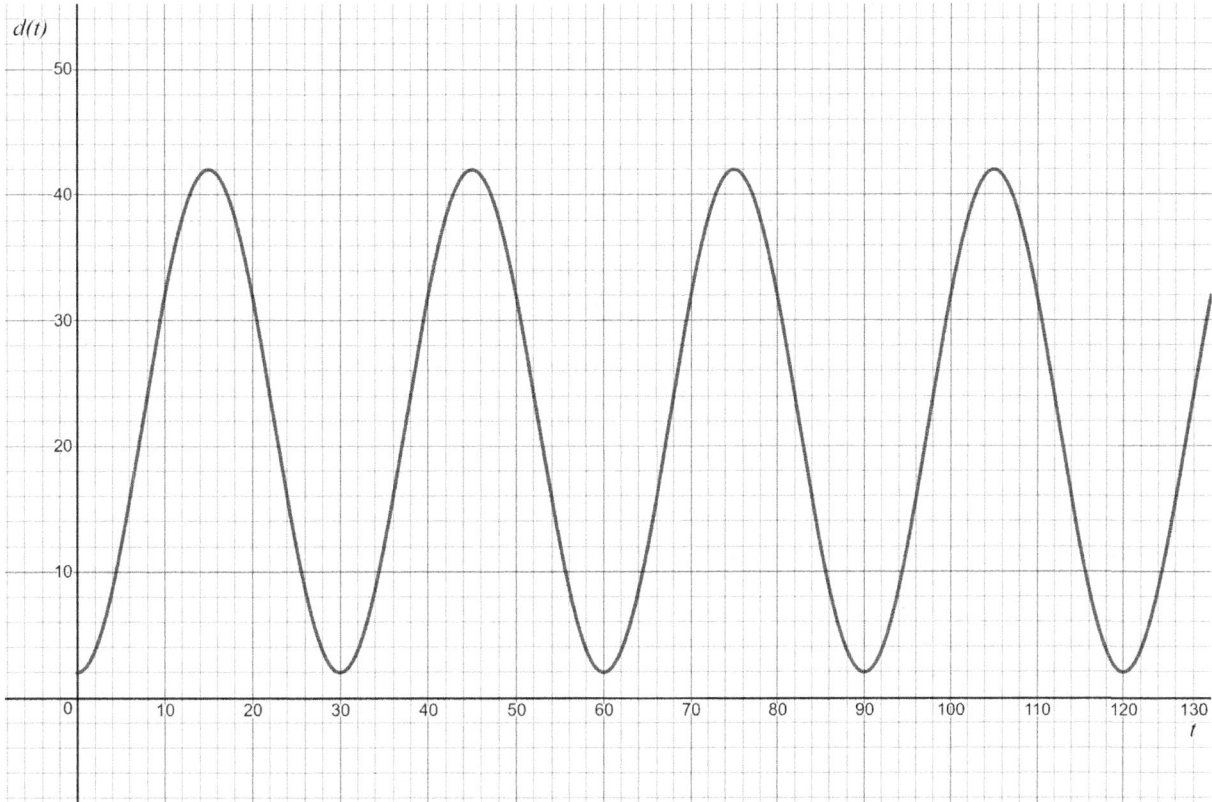

(a) What is the maximum height reached by the rider of the Ferris wheel? What are the first two times this maximum occurs?

(b) What is the minimum height reached by the rider of the Ferris wheel? What are the first two times this minimum occurs?

(c) How long does it take the rider to travel make one complete revolution on the Ferris wheel?

(d) Find four times the rider is at a height of 32 feet?

(e) During the first 30 seconds, when is height of the rider at least 32 feet?

(f) Solve $d(t) < 32$ on the interval $[0, 30]$.

1.3   Modeling with Linear Functions

Goals:

- L: Be able to solve a linear equation.

- L: Be able to model a situation with appropriate linear equation(s) and interpret the solution.

- L: Be able to determine the slope or equation of a linear function given its graph or a table of values.

- F: Be able to determine inputs or outputs from a function table.

- F: Be able to determine inputs or outputs from a function graph.

Problem 17. The table below shows a number pattern. Fill in the blanks in the table, and make a graph of the values in the table. Then, write a function equation for f.

n	$f(n)$
1	1
2	3
3	5
4	7
5	
6	
20	

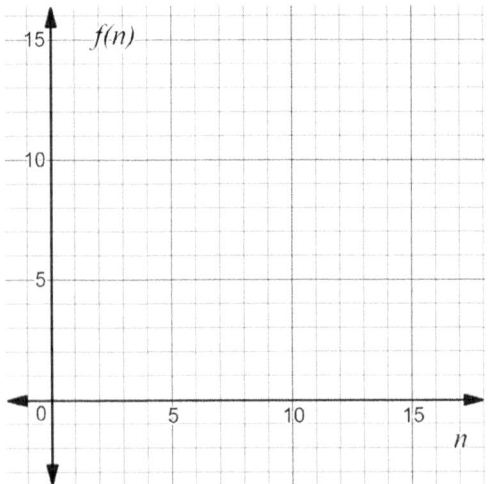

Investigation 18. Nancy is a landscape artist. She specializes in square ponds surrounded by hand-painted tiles. Customers can order a pond in any size square, starting with sides measuring 2 feet, and available in one-foot increments thereafter (sides of 3 feet, 4 feet, 5 feet, etc.). The tiles are 1-foot square, and are placed edge-to-edge along the entire outer perimeter of the pond. An example pond is shown below.

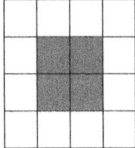

1. Make a table showing the number of border tiles needed for different pond sizes, from 2-foot sides up through 6-foot sides.

2. If the side length of the pond is increased by 1 foot, how many more border tiles are needed? Explain why the number of tiles is increasing according to this pattern.

3. How many border tiles are needed for a pond with sides of length 12 feet?

4. If Nancy orders 64 border tiles for an upcoming job, how large is the pond the customer wants?

5. Describe in words how to find the number of tiles needed for the border of a pond.

6. Write an equation for the number of tiles as a function of the length of one side of the pond.

Problem 19. A local doughnut shop charges $9 for the first dozen doughnuts purchased, and $0.50 for each additional individual doughnut purchased.

1. Write a linear function for $C(d)$, the cost of d doughnuts.

2. For what values of d does your equation from Part 1 make sense?

3. How much will 20 doughnuts cost?

4. Your club is having a fundraiser reselling doughnuts from this shop. You were given $20 to buy doughnuts to resell. How many doughnuts can you buy?

5. Using your answer from part 4, what is the average cost of each doughnut?

6. Your club wants to sell the doughnuts for $1 and make a profit of at least $0.45 on each doughnut. What is the minimum number of doughnuts you need to buy (and sell) in order for this to happen?

Problem 20. Stephanie is planting her summer garden. She purchased a sunflower plant from a nursery and planted it when it was 6 cm tall. After she planted it, the sunflower plant grew at a constant rate of 3 cm per day.

1. Write a function, $H(t)$ that gives the height of the sunflower plant as a function of t, the number of days since the plant was planted.

2. A worker at the nursery told Stephanie that her sunflower would reach its maximum height about 40 days after it was planted. What will the height be at that time?

3. How many days after planting would the sunflower be half its maximum height?

4. Consider your function from part 1. What domain and range make sense in the context of this problem?

Problem 21. Isabella is considering renting a car for one day to drive to an event in another city. Company A charges \$20 per day plus \$0.25 per mile, and Company B charges \$25 per day plus \$0.15 per mile.

1. Write a linear function equation that models the one day cost of renting a car from Company A, C_A, as function of m, the number of miles driven.

2. Write a linear function equation that models the one day cost of renting a car from Company B, C_B, as function of m, the number of miles driven.

3. Graph C_A and C_B on the axes below. Use the graph to estimate the number of miles for which the cost for Company A is the same as the cost for Company B.

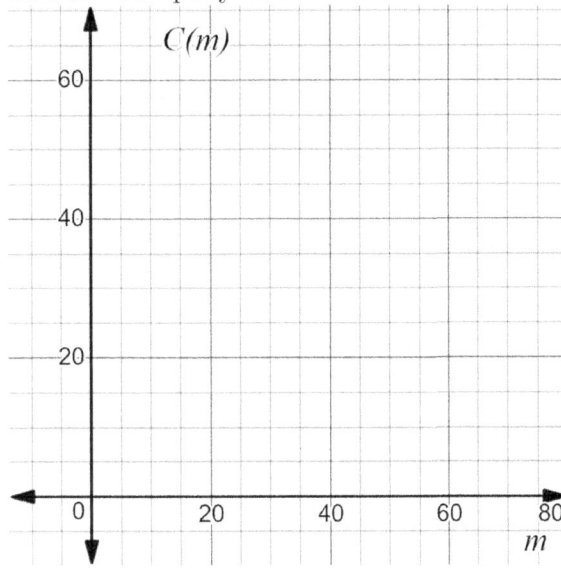

4. Use algebra to find the number of miles for which the cost for Company A is the same as the cost for Company B.

5. If the event Isabella is going to is a 60 mile round trip drive, which rental company should she choose? How much will she save by choosing that company?

Notes

1.3 Exercises

1. Refer to Figure 1.2, Triangle Pattern. Assume that each edge of the triangle measures 1 cm. Assume that one triangle is added to one figure to get the next figure. Let P be the function describing the perimeter as a function of the figure number.

   (a) Draw the next two figures in the pattern.
   (b) Make a table with columns for n and $P(n)$ for $1 \le n \le 5$.
   (c) Make a graph of the values in your table.
   (d) Write an equation for $P(n)$.
   (e) Find $P(12)$.
   (f) How many triangles are in a figure with a perimeter of 18?

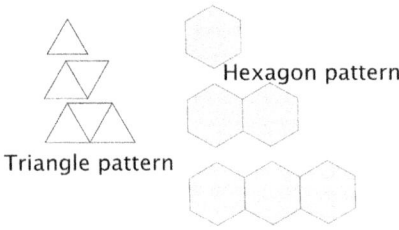

Hexagon pattern

Triangle pattern

Figure 1.1: More Patterns

2. Refer to Figure 1.2, Hexagon Pattern. Assume that each edge of the hexagon measures 1 cm. Assume that one hexagon is added to one figure to get the next figure. Let H be the function describing the perimeter as a function of the figure number.

   (a) Draw the next two figures in the pattern.
   (b) Make a table with columns for n and $H(n)$ for $1 \le n \le 5$.
   (c) Make a graph of the values in your table.
   (d) Write an equation for $H(n)$.
   (e) Find $H(15)$.
   (f) Solve $H(n) = 38$.

3. MovieTicket is offering a discount plan where if you subscribe to the plan for $20 per month, you can purchase as many movie tickets as you want for $5 instead of the regular price of $11. How many movies must you see a month to make it worth signing up for the plan?

4. A sandwich shop charges $7 for a foot-long sandwich with one protein and unlimited veggies. You can add additional proteins for $1.50 per protein.

   (a) Write a linear function equation that models S, the cost of a sandwich, as a function of t, the number of toppings on the sandwich.

   (b) The 'Monster Meat' sandwich has all 7 of the proteins that the sandwich shop has available. How much will this cost?

(c) If your sandwich cost $13, how many proteins did it have?

(d) Consider your linear function from part 4a. What domain and range make sense in the context of this problem?

5. Sally needs to choose between two internet plans. Under the first plan Verizon will sell her a DSL modem for $17.99, then she must pay $12.99 per month. Under the second plan, ATT will give her a modem for free, but she must pay $14.99 per month.

  (a) When does the Verizon plan become the better deal (how many months)?

  (b) Describe at least one other possible solution method for part (5a) besides the one you used.

6. Barbara is visiting another county and wants to know the sales tax rate. She just bought a pair of shorts for $22 and paid $1.87 in sales tax.

  (a) What tax will she pay if she buys sunblock for $8?

  (b) What is the sales tax rate?

  (c) Write a function equation that computes the tax as a function of the item price.

7. The U.S. is nearly the last country to use the English system of measurement, which includes the Fahrenheit scale for temperature instead of the Celsius scale. There are two conversion formulas, one from Celsius to Fahrenheit, and one from Fahrenheit to Celsius. One of the two formulas is $F = \frac{9}{5}C + 32$, where F is the temperature in degrees Fahrenheit, and C is the temperature in degrees Celsius.

  (a) Explain the significance of the F-intercept in the given formula.

  (b) Find the conversion formula for Fahrenheit to Celsius.

  (c) When traveling abroad, it can be useful to know the formula, but it is not necessary to remember the formula. One way to generate the formula is by remembering two temperatures on both scales. Why is it enough to remember just two temperatures in both scales?

  (d) In fact, this author remembers three temperatures: $0^{o}C$ is $32^{o}F$, $20^{o}C$ is $68^{o}F$, and $100^{o}C$ is $212^{o}F$. Choose a pair of temperatures and use them to recover the conversion formula from Celsius to Fahrenheit.

Temp (^{o}F)	59	68	77	86	95
Temp (^{o}C)		20			

Table 1.1: Temperature conversions

8. The graph of f is shown below.

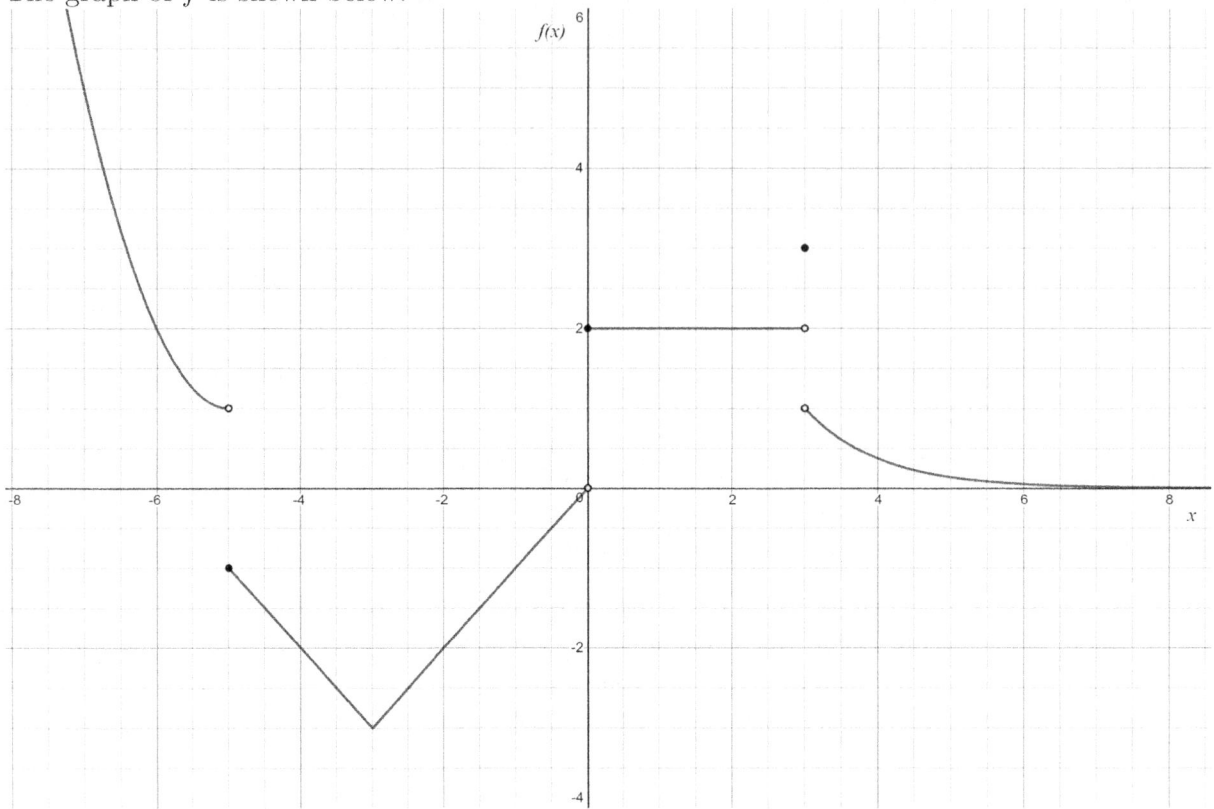

(a) Evaluate $f(-4)$

(b) Evaluate $f(-5)$

(c) Evaluate $f(3)$

(d) Solve $f(x) = 5$

(e) Solve $f(x) = -2$

(f) Solve $f(x) = 2$

(g) Solve $f(x) = 0$

(h) Solve $f(x) > 5$

(i) Solve $f(x) \leq -2$

1.4 Linear Parametric Equations

Parametric equations are often used to keep track of the position of an object in motion. In this case, rather than having y as a function of x, usually we have x as a function of t (time), and y as a function of t.

Goals:

- F: Be able to draw a diagram incorporating all of the important information in a given situation, and impose coordinates on the diagram.

- L: Be able to model a situation involving linear motion with appropriate parametric equation(s) and interpret the solution.

Investigation 22. The science quad at Julian's school measures 120 ft by 90 ft. Julian begins walking at a constant rate diagonally from the Northwest corner of the quad to the Southeast corner.

1. The URL `https://tinyurl.com/153ParaActivity` shows Julian's path across the quad with coordinates imposed. Sketch his path below.

2. What is the length of Julian's path across the quad?

3. If Julian is walking at a speed of 5 feet per second, how long does it take him to walk the diagonal of the quad? In Desmos, move the slider for a, where a represents the number of seconds since Julian started walking, to see how Julian moves along his path. Label the starting and ending time on your graph.

4. What are Julian's coordinates at time $a = 0$? What are Julian's coordinates at time $a = 30$? Label these times on your graph.

5. What are Julian's coordinates at time $a = 20$?

6. At what time is Julian at the point $(20, 75)$?

7. So far this semester, we have worked with functions that are of the form $y = f(x)$, where y depends on x. Using this representation, Julian's path can be represented by the function $y = f(x) = -\frac{3}{4}x + 90$, $0 \le x \le 120$.

   Julian's path can be represented by the parametric equations $(4t, -3t + 90)$ where the $x-$coordinate of his position is given by $x(t) = 4t$ and the y-coordinate of his position is given by $y(t) = -3t + 90$, $0 \le t \le 30$. What additional information does this representation of his path have that the first representation does not?

Example 23. A woman is walking in a park. She begins at a point 30 meters west of the northwest corner of the playground, and walks in a straight line to a pond 60 meters north of the same corner of the playground.

1. Draw a diagram showing the woman's path and impose coordinates on the diagram.

2. Write an equation for the line that describes the woman's path.

3. It takes the woman 20 seconds to walk to the pond. Write linear parametric equations to describe her position at time t seconds.

4. It takes a different woman 30 seconds to walk the same path, but in the opposite direction. Write linear parametric equations to describe the second woman's position at time t.

The steps used to solve Example 23 are outlined below:

To solve part 1, we draw Figure 1.3, where W is the woman's starting position, and P is the pond. (Other diagrams are possible.) To solve 2, find the equation passing through the two points $(-30, 0)$ and $(0, 60)$. The slope is $\frac{60-0}{0-(-30)} = 2$, and the y-intercept is 60, so the equation is $y = 2x + 60$.

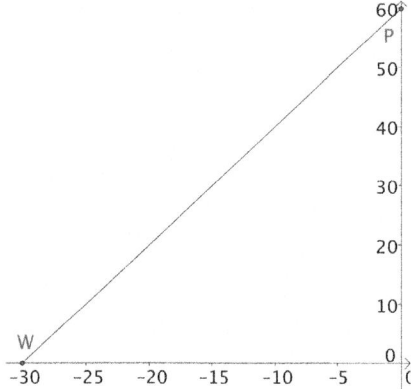

Figure 1.2: A walk in the park

To solve 3, we are looking for two equations, $x(t)$ and $y(t)$.

Step 1: Get values for t, x, and y at two points in time. In this problem, we know that at $t = 0$, we are at point W, so $x = -30$ and $y = 0$, and at $t = 20$ seconds, we are at point P, where $x = 0$ and $y = 60$.

Step 2: Get the $x(t)$ linear equation. We are going to find the equation of a line where we treat t as the independent value and x as the dependent value. Notice that the slope will be $\frac{\triangle x}{\triangle t}$. So we have $\frac{0-(-30)}{20-0} = \frac{3}{2}$, and since at $t = 0$, $x = -30$, then our equation is $x(t) = \frac{3}{2}t - 30$.

Step 3: Get the $y(t)$ linear equation. We are going to find the equation of a line where we treat t as the independent value and y as the dependent value. Notice that the slope will be $\frac{\triangle y}{\triangle t}$. We have $\frac{60-0}{20-0} = 3$. At $t = 0$, $y = 0$, so our equation is $y(t) = 3t$.

Step 4: Verify the solution. On Desmos, type in $(\frac{3}{2}t - 30, 3t)$, and set $0 \le t \le 20$. You should see the path from Figure 1.3.

Problem 24. Alberto is at the beach, and is hungry for a hot dog. Currently, he is standing at the water's edge, 100 meters west of the boardwalk. Directly in front of him on the boardwalk, the nearest food vendor is his friend Carmen, who sells pizza. A hot dog vendor is 40 meters south of Carmen's pizza stand. Alberto plans to walk directly to the hot dog vendor.

1. Draw a diagram showing Alberto's path and impose coordinates on the diagram.

2. Write an equation for the line that describes the Alberto's path.

3. Write parametric equations to describe Alberto's position at time t, if it takes him 90 seconds to walk from the water to the hot dog vendor.

Notes

1.4 Exercises

1. Write parametric equations for a point traveling along the line $y = 2x - 6$, such that at $t = 0$ the point is at the x-intercept, and at $t = 1$ the point is at the y-intercept.

2. A volleyball court measures 18 meters by 9 meters. Jane takes 15 seconds to walk from one corner of the court to the diagonally opposite corner.

    (a) Draw a diagram of the court and Jane's path across the court, and impose coordinates on the diagram.

    (b) Write parametric equations to describe Jane's path.

3. An object is moving along a line in the xy-plane so that at time $t = 0$, it is at the point $(8, 16)$ and at $t = 4$, it is at the point $(0, -20)$.

    (a) Draw a diagram of the xy-plane showing the object's position at time $t = 0$ and $t = 4$.

    (b) Write parametric equations $(U(t), V(t))$ to describe the position of the object at time t.

4. Erwin is at an outdoor market. He walks in a straight line from a fruit stand at a point 65 feet due West of a fountain to a florist at a point 420 feet due North of the fountain. He walks at a constant speed of 5 feet per second.

    (a) Draw a diagram showing the location of the fountain, the fruit stand, and the florist, and impose coordinates on the diagram.

    (b) Write an equation describing Erwin's path through the market.

    (c) Write parametric equations for Erwin's position t seconds after he begins walking.

5. The graph below shows the period of pendulum T (the time in seconds of one complete oscillation) as a function of the length of the pendulum, l (in meters).

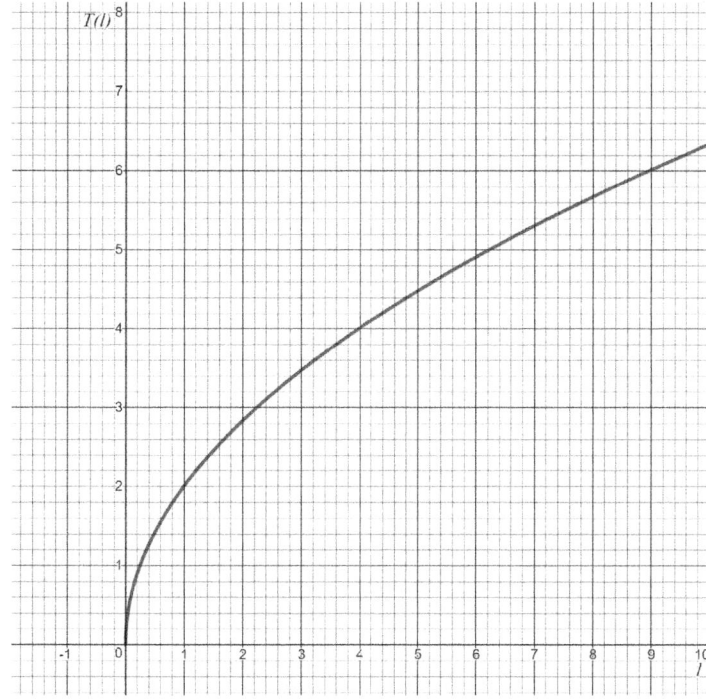

(a) Use the graph above to find the period (in seconds) of a pendulum of length 3 m.

(b) Use the graph above to find the length of a pendulum that has a period of 2.5 seconds.

(c) If the relationship between the length and period of a pendulum is given by the formula $T(l) = 2.006l^{\frac{1}{2}}$, use the formula to find the period (in seconds) of a pendulum of length 3 m.

(d) Using the equation in part 5c find the length of a pendulum that has a period of 2.5 seconds.

6. Mr. and Mrs. Jones both teach mathematics. By 1999, Mrs. Jones had taught 300 students, and has an average of 150 students each year. Mr. Jones began teaching his own classes in 1999, and has an average of 100 students each year.

   (a) Write a function equation to estimate the total number of students Mrs. Jones has taught as a function of t, the number of years since 1999.

   (b) Write a function equation to estimate the total number of students Mr. Jones has taught as a function of t, the number of years since 1999.

   (c) In what year has Mrs. Jones taught 1000 *more* students than Mr. Jones?

7. Evaluate each function for the indicated value or expression.

   (a) $r(t) = \sqrt{t^2 - 1} + 1$, $r(-2)$

   (b) $f(x) = x^3 - 2x$, $f(3)$

   (c) $c(x) = x^2 - 1$, $c(x + 2)$

   (d) $d(t) = 4t^{\frac{2}{3}}$, $d(8)$

8. Let $f(x) = -x - 2$, $g(x) = 4x + 1$, $h(x) = 2(x - 3) + 1$

   (a) Evaluate $h(-2)$

   (b) Evaluate $f(0)$

   (c) Solve $g(x) = -2$

   (d) Solve $f(x) = g(x)$

   (e) Solve $h(x) \leq -1$

   (f) Solve $f(x) > 0$

   (g) Solve $f(x) < h(x)$

Chapter 2

Quadratic Functions

2.1 Solving Quadratic Equations

Goals:

- Q: Be able to determine the vertex and equation of a quadratic function given its graph.

- F: Be able to give the solution to an inequality or set of inequalities using proper mathematical notation.

- F: Be able to determine the domain or range of a function given as an equation or a graph.

Definition 25. A **quadratic function** is a function that can be written as

$$f(x) = ax^2 + bx + c$$

where $a \neq 0$. The shape of the graph of a quadratic function is a **parabola**. The minimum or maximum value of a quadratic function is at the **vertex** of its associated parabola.

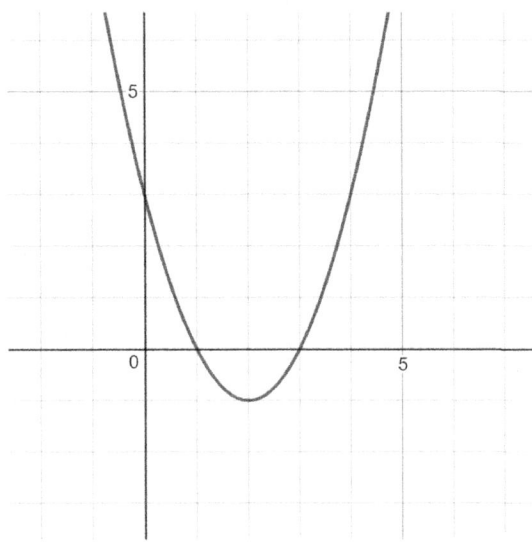

Problem 26. Decide if each statement below is always, sometimes or never true.

1. A parabola has two $x-$intercepts. Always/Sometimes/Never

2. A parabola has exactly one $y-$intercept. Always/Sometimes/Never

3. A quadratic function has a minimum value. Always/Sometimes/Never

4. A quadratic function has three terms. Always/Sometimes/Never

Solving Quadratic Equations by Factoring

Quadratic equations can be solved in several ways: by factoring, by using the quadratic formula, by completing the square or by using a graph. We will first consider solving quadratic equations by factoring, and see how this compares to solving quadratic equations using a graph.

Definition 27. To solve quadratic equations by factoring, we will use the **zero-product property**, which says that if $ab = 0$, then $a = 0$ or $b = 0$.

Example 28. Solve each equation using the zero-product property.

1. $(2x-1)(x+3) = 0$
2. $x(x-1)(x+2) = 0$
3. $5(x-2)^2 = 0$

Problem 29. Solve each equation below by factoring.

1. $x^2 - 5x - 6 = 0$
2. $a^2 + 2 = -3a$
3. $r^2 - 16 = 0$

4. $8n^2 + n = 7$
5. $v^2 = 5v$
6. $-4x^2 - 20x + 24 = 0$

Many quadratic expressions you will encounter in pre-calculus and calculus will be easy to factor using trial and error. If you need to factor a quadratic expression where the coefficients have many factors, trial and error can be time consuming. In these cases, it is useful to have a general method of factoring that relies less on trial and error.

Problem 30. Solve each equation below by factoring.

1. $12x^2 + 8x - 15 = 0$
2. $12x^2 - 32x = -5$

Definition 31. The following **special products** come up so often that it is worth memorizing them.

- Difference of squares: $a^2 - b^2 = (a - b)(a + b)$

- Perfect square (sum): $a^2 + 2ab + b^2 = (a + b)^2$

- Perfect square (difference): $a^2 - 2ab + b^2 = (a - b)^2$

Problem 32. Solve each equation below by factoring.

1. $x^2 = 100$
2. $x^2 - 6x + 9 = 0$
3. $25x^2 + 10x = -1$

Some equations can be solved by factoring, even if they are not quadratic equations of the form $ax^2 + bx + c = 0$.

Example 33. Solve by factoring.

1. $n^4 - 4n^2 + 3 = 0$

2. $2f^4 = 6f^3 + 8f^2$

3. $14r^3 + 7r^2 = 7r$

4. $0 = x^4 - 2x^2 + 1$

Not every quadratic equation can be solved by factoring. One method of solving any quadratic equation is using the quadratic formula.

Definition 34. The **quadratic formula** is

$$x = \frac{-b \pm \sqrt{b^2 - 4ac}}{2a}$$

where a, b, and c are coefficients of the quadratic equation $ax^2 + bx + c = 0$.

Example 35. Solve $2x^2 - x - 4 = 3$

Problem 36. Find the x-intercepts of each quadratic function. Use Desmos to sketch a graph of the function and label the x-intercepts.

1. $f(x) = -3x^2 - x + 3$

2. $g(x) = 2x^2 - 4x + 5$

3. $k(x) = 4x^2 - 12x + 9$

4. From your answers to part 1 through 3 above, explain how you can tell if a quadratic equation has no solutions, one solution or two solutions.

Example 37. Let $f(x) = -x^2 + 2x + 3$

1. Find the $x-$ and $y-$intercepts of the graph of f.

2. Find the vertex of the graph of f.

3. Use parts 1 and 2 to graph f. Use Desmos to check your answer.

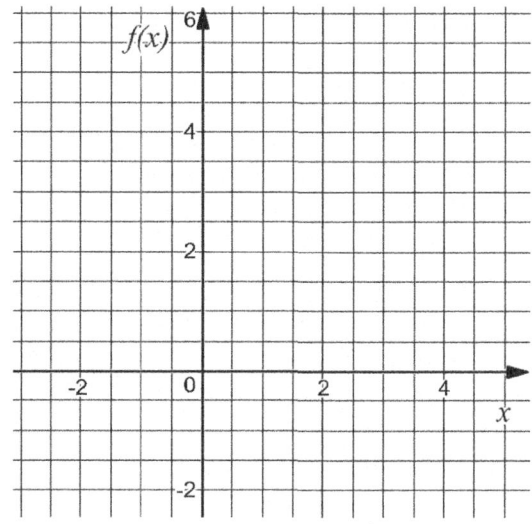

4. Find all the values of x where $f(x) = 3$.

5. Solve $f(x) \leq 3$.

6. Solve $f(x) = 1$ for x. Give your answer as an exact value and a decimal approximation.

Problem 38. Let $g(x) = 2x^2 - 12x + 10$.

1. Find the $x-$ and $y-$intercepts of the graph of g.

2. Find the vertex of the graph of g.

3. Use parts 1 and 2 to graph g. Use Desmos to check your answer.

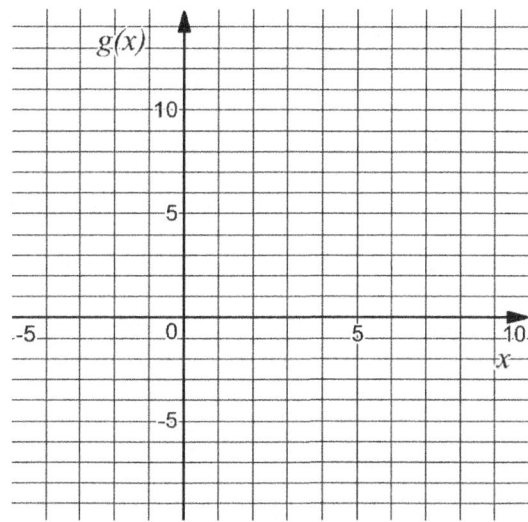

4. Solve $g(x) = -6$ for x.

5. Solve $g(x) > -6$.

Notes

2.1 Exercises

1. Refer to $g(t) = t^2 - 8t + 15$.

   (a) Graph the function g on the domain $-6 \le t \le 6$. Be sure to label the vertex and any intercept(s).
   (b) Solve the inequality $g(t) > 3$.
   (c) Solve the inequality $g(t) < t^2 - 1$.
   (d) What is the range of g on the domain of all real numbers?

2. Solve each equation.

   (a) $a^2 - 100 = 300$
   (b) $2y^2 - 2y = -5y$
   (c) $-3x + 2 = -x^2$
   (d) $n^4 - 4n^2 = -4$
   (e) $-4 = 9x^2 + 12x$
   (f) $5t^2 + t - 2 = 0$
   (g) $\frac{1}{2}w^2 + \frac{2}{3}w - \frac{1}{3} = 0$
   (h) $(2z + 1)^2 - 3 = -2(2z + 1)$
   (i) $5x^2 - 3x = -2x^3$
   (j) $2b^2 - 14b + 24 = 0$

3. The graph below shows the number of students, N, in Math Club t years after 1990, when the club started.

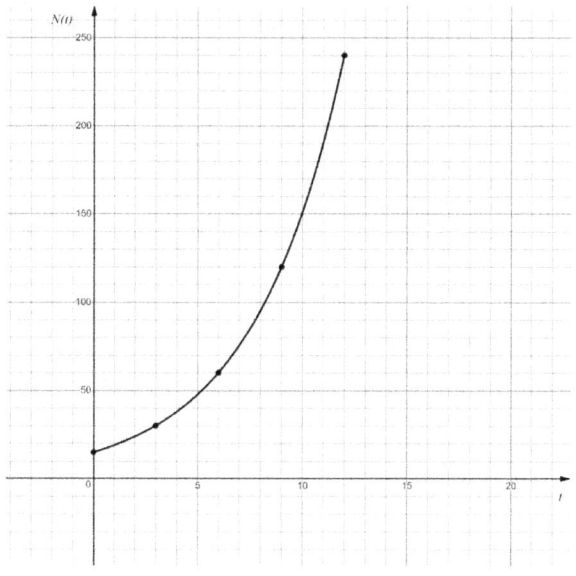

   (a) How many students were in Math Club in 1999?

   (b) When were there approximately 50 students in Math Club?

(c) How many students were in Math Club when it started in 1990?

(d) When will the number of students in Math Club reach four times the number in 1990?

(e) How is the number of students joining Math Club changing over time?

4. Let $g(x) = 3x^2 + 2x - 1$.

  (a) Find the $x-$ and vertical intercepts of the graph of g.

  (b) Find the vertex of the graph of g.

  (c) Use parts 4a and 4b to graph g. Use Desmos to check your answer.

  (d) Solve algebraically: $g(x) = 7$. Use Desmos to check your answer.

  (e) Solve algebraically: $g(x) = 5$. Use Desmos to check your answer.

  (f) Solve $g(x) > -1$.

5. According to the U.S. Library of Congress (https://www.loc.gov/rr/scitech/mysteries/cricket.html), you can estimate the temperature by listening to the chirping of a cricket. At 70 °F a cricket will chirp 113 times per minute, and at 80 °F a cricket will chirp 173 times per minute.

  (a) Write a linear function for $T(c)$, the temperature when a cricket makes c chirps per minute.

  (b) What is the temperature if a cricket chirps 100 time per minute?

  (c) How many times a minute will a cricket chirp if it is 52 °F?

  (d) In part 5a you found the slope of the linear function describing the relationship between the number of times a cricket chirps per minute and the temperature. What does this slope represent?

  (e) In part 5a you found the vertical intercept of the linear function describing the relationship between the number of times a cricket chirps per minute and the temperature. What does this vertical intercept represent?

6. Table 2.1 shows a linear pattern. Make a graph of the values in the table. Then, write a function equation that gives the output value corresponding to any input.

n	$g(n)$
1	17
2	14
3	11
4	8

Table 2.1: Patterns

7. A basketball court is 94 feet by 50 feet. Lisa takes 20 seconds to walk from one corner of the court to the diagonally opposite corner.

  (a) Draw a diagram of the court and Lisa's path, and impose coordinates on the diagram.

  (b) Write parametric equations to describe Lisa's path.

2.2 Quadratic Functions

Goals:

- Q: Be able to solve a quadratic equation.

- Q: Be able to determine the vertex and equation of a quadratic function given its graph.

- F: Be able to give the solution to an inequality or set of inequalities using proper mathematical notation.

- F: Be able to determine the domain or range of a function given as an equation or a graph.

Definition 39. In Section 2.1, we saw that a quadratic function can be written as $f(x) = ax^2 + bx + c$. A quadratic function can also be written in **vertex form** as $f(x) = a(x - h)^2 + k$, where $a \neq 0$ and h and k are real numbers and x is a variable.

Investigation 40. In Desmos, graph $f(x)$ and create sliders for a, h and k. What role do each of these constants play? Include sketches and verbal descriptions to help explain the role of each constant.

Remark 41. Sometimes we will have the graph of the quadratic function. Other times we have the equation for the quadratic function, but we may have to do some work to put it the form in Definition 39. As we've seen with linear functions, the names of the variables don't matter. Instead of x you might see t or another variable, and instead of f the function may be named with a different letter.

Example 42. Refer to the graph of k in Figure 2.1.

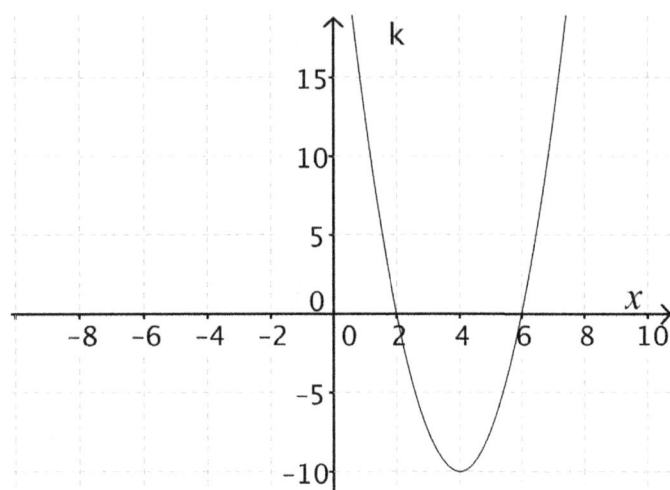

Figure 2.1: Graph of k

1. What are the coordinates of the vertex of the graph of k?

2. Write a function equation for $k(x)$.

3. Solve $k(x) = 15$ for x.

4. Solve the inequality $k(x) \leq 15$.

5. What is the range of k on the domain of all real numbers? What is the range of k on the domain $0 \leq x \leq 6$?

Problem 43. Refer to the graph of q in Figure 2.2.

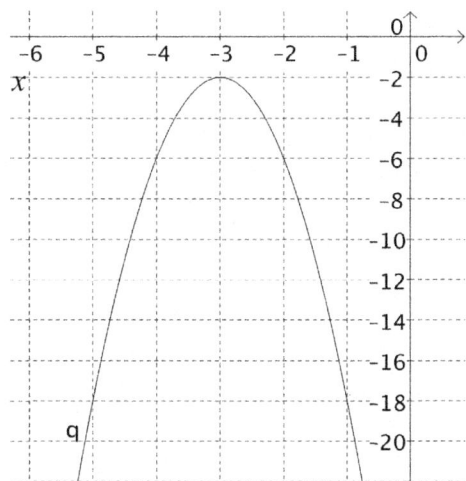

Figure 2.2: Graph of q

1. Evaluate $q(-2)$.

2. What are the coordinates of the vertex of $q(x)$?

3. Write a function equation for $q(x)$.

4. Use your equation from part 3 to find $q(-5)$. Verify the point on your graph.

5. Solve the inequality $q(x) \leq -6$.

6. Solve the inequality $q(x) > 2x - 12$.

7. What is the range of q on the domain of all real numbers?

We have seen how to use the graph of a quadratic function to write an equation of the function in vertex form. We can also begin with a quadratic function written as $f(x) = ax^2 + bx + c$ and write it in vertex form.

Example 44. Let $f(x) = x^2 - 6x + 11$.

1. Write f in vertex form.

2. Graph f on the axes below.

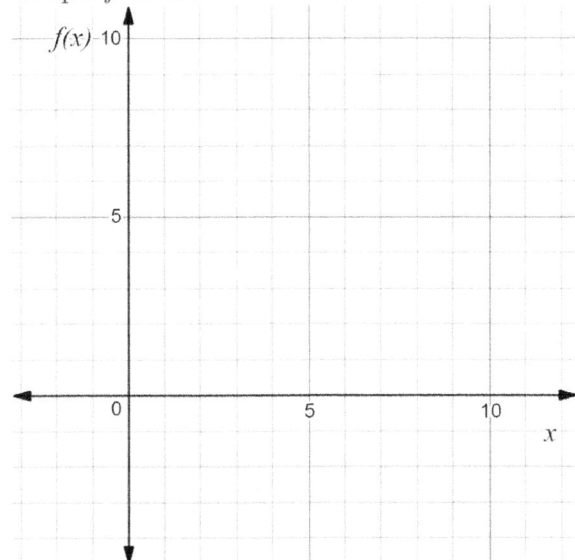

3. Solve $f(x) = 0$.

4. Solve $f(x) = 2$.

5. What is the domain of f? What is the range of f?

6. What is the minimum value of f? For what value of x does this minimum occur?

Problem 45. Let $g(x) = -2x^2 + 4x + 1$.

1. Write g in vertex form.

2. Graph g on the axes below.

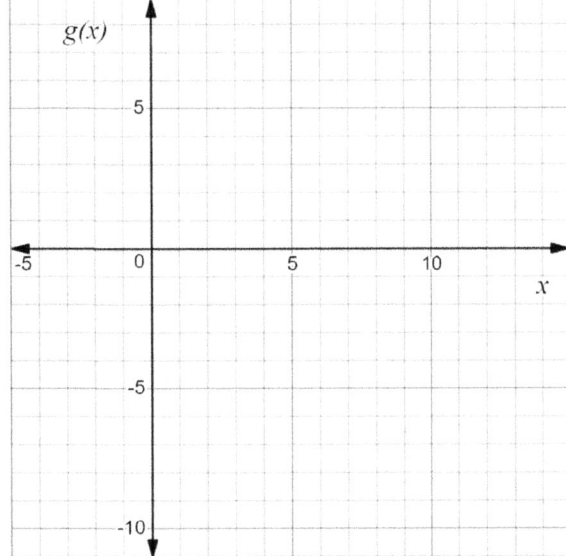

3. Solve $g(x) = 0$.

4. Solve $g(x) = -4$.

5. What is the domain of g? What is the range of g?

6. What is the maximum value of g? For what value of x does this maximum occur?

Example 46. Solve each inequality algebraically. Verify your answer using Desmos.

1. $y^2 - 6y - 16 > 0$

2. $-2x^2 - 5x \leq -3$

Problem 47. Solve each inequality algebraically. Verify your answer using Desmos.

1. $(w - 4)(w - 5) < 2$

2. $-x^2 - 6x + 55 \geq 0$

Notes

2.2 Exercises

1. Refer to the graph of v in Figure 2.3.

    (a) What are the coordinates of the vertex of the graph of v?
    (b) Write a function equation for $v(t)$.
    (c) Solve the inequality $v(t) \geq -12$.
    (d) What is the range of v on the domain of all real numbers?

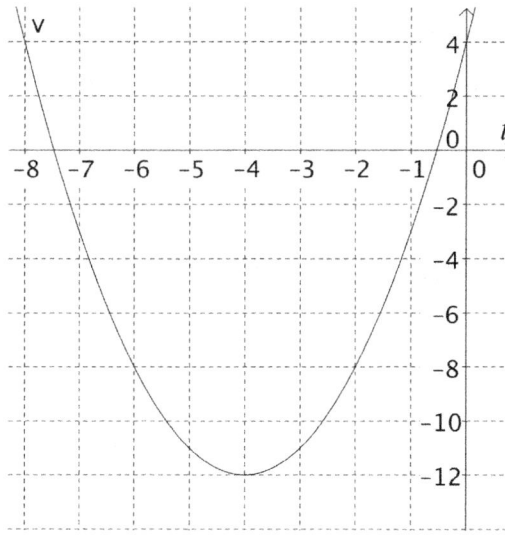

Figure 2.3: Graph of v

2. Refer to the graph of g below.

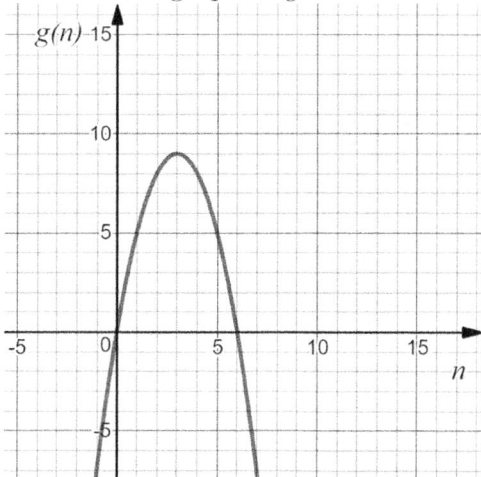

    (a) Evaluate $g(5)$.

    (b) Solve $g(n) = 9$.

    (c) Write an equation for $g(n)$.

    (d) Use your equation from part 2c to evaluate $g(10)$.

(e) Use your equation from part 2c to solve $g(n) = -8$. Use Desmos to check your answer.

(f) Use your equation from part 2c to solve $g(n) < 8$.

3. Let $k(x) = x^2 - 8x + 10$

    (a) Write $k(x)$ in vertex form.

    (b) Sketch the graph of $k(x)$.

    (c) Solve $k(x) = 0$

    (d) Solve $k(x) = 6$

    (e) Solve $k(x) < -2$

4. Let $q(x) = 2x^2 - 8x + 7$

    (a) Write $q(x)$ in vertex form.

    (b) Sketch the graph of $q(x)$.

    (c) Solve $q(x) = 0$

    (d) Solve $q(x) = 3$

    (e) Solve $q(x) \geq -1$

5. Solve each inequality algebraically. Check your answers on Desmos.

    (a) $x^2 > x + 2$

    (b) $x^2 + 9x + 13 \geq -7$

    (c) $4x^2 + 8 \leq 33$

    (d) $-x^2 + 6x < 8$

6. Graph the path traveled by a point described by the parametric equations $x(t) = 3 + 4t$, $y(t) = 8 - 2t$ on the domain $-2 \leq t \leq 5$. Label the location of the point at $t = 0$.

7. Refer to the graph of h below.

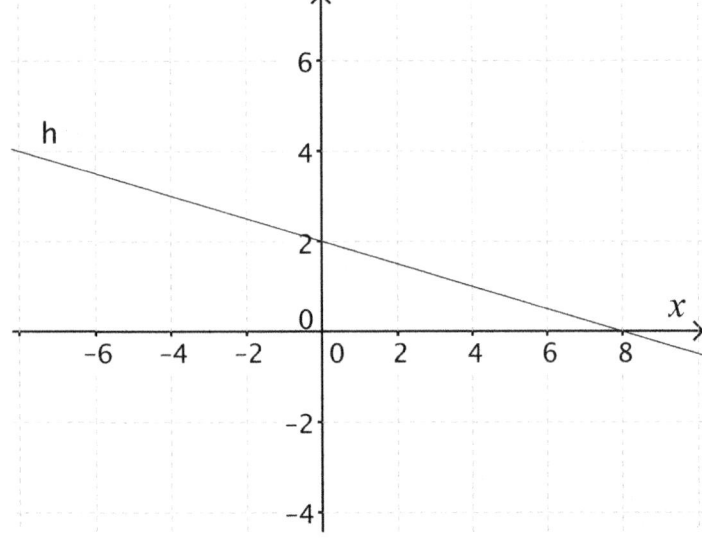

(a) What is $h(4)$?

(b) Write a function equation for $h(x)$.

(c) Use your function equation to evaluate $h(4)$. Does this match with your answer from part 7a?

8. The graph below shows the height, H (in cm), of a tomato plant t days after it is planted outdoors.

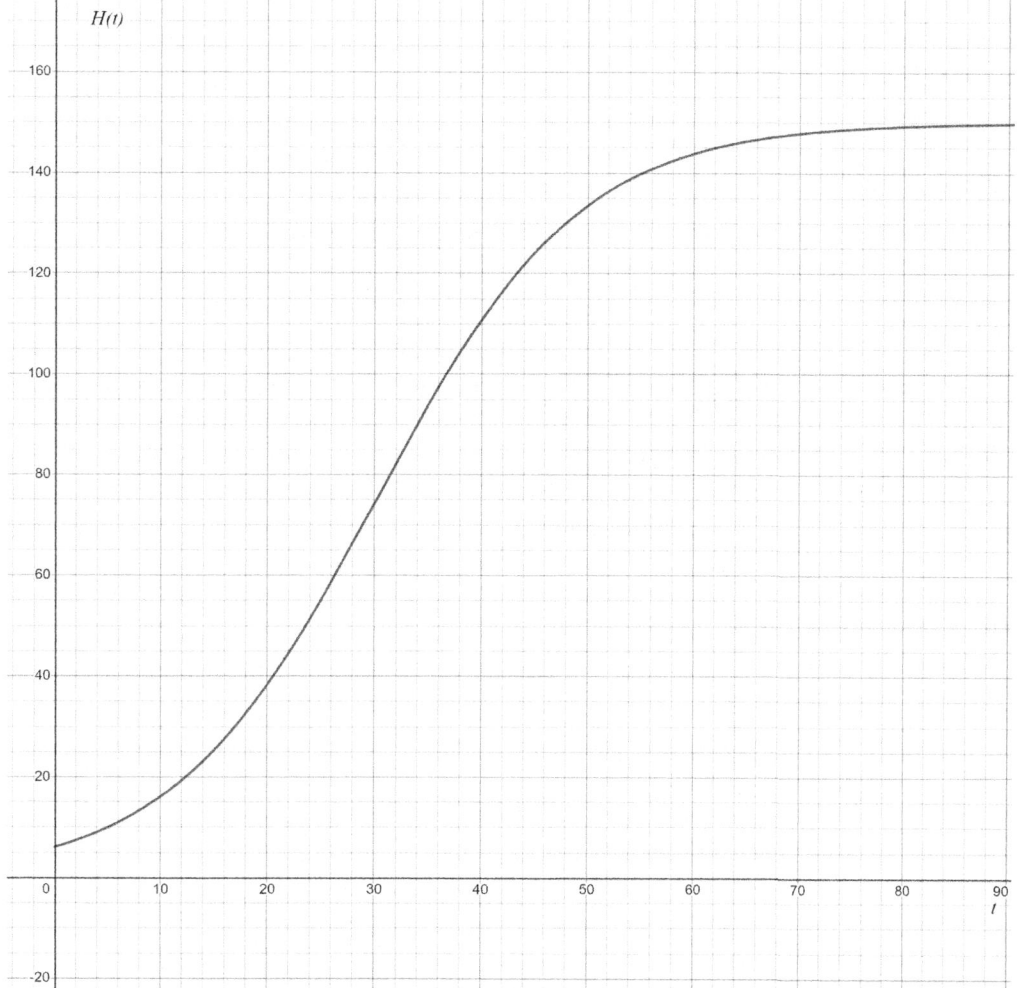

(a) What is the height of the plant on the day it is planted outdoors?

(b) How tall is the plant after 30 days?

(c) How many days does it take for the plant to reach 80 cm?

(d) Where is the height of the plant increasing the slowest? Where is the height of the plant increasing the fastest?

2.3 Quadratic Modeling

Quadratic functions are useful as mathematical models. We will use them to model objects experiencing the force of gravity, or for modeling income when demand is a linear function of the price.

Goals:

- Q: Be able to model problem situations with an appropriate quadratic function equation(s) and interpret the solution(s).

- Q: Be able to model motion of objects falling with the force of gravity with appropriate quadratic equation(s) and interpret the solution.

- Q: Be able to determine the equation of a quadratic function given its graph.

- Q: Be able to determine and interpret the vertex of a quadratic function given an equation or context.

Problem 48. A flower pot falls from the ledge of a balcony on a high-rise building. An object experiencing the force of gravity can be modeled by the equation $h(t) = -16t^2 + vt + c$, where t is the time in seconds, $h(t)$ is the height in feet, c is the initial height of the object, and v is the initial velocity of the object. **Note:** This model applies to any object experiencing the force of gravity, where the measurement is in English units (feet). There is another version using metric units (meters). This model also assumes air resistance is negligible, so it doesn't work well for things like feathers and parachutes where air resistance is strong.

1. At the instant the pot begins to fall, what is its initial velocity?

2. Suppose it takes 3 seconds for the pot to hit the ground. How high was the balcony?

3. Write a formula for h that models the height of the pot at time t.

4. What is the height of the pot at time $t = 1.5$ seconds?

5. When will the height of the pot be 100 feet?

Problem 49. A ball is thrown upward from a height of 5 feet. It takes the ball 3.25 seconds to hit the ground.

1. How fast is the ball being thrown at the instant it is released?

2. Write a formula for h that models the height of the ball at time t.

3. What is the maximum height of the ball? At what time does the ball reach this height?

4. What is the height of the ball at $t = 1$ second?

5. At what time(s) is the height of the ball 13 feet?

6. When is the height of the ball at least 13 feet?

Problem 50. Esperanza owns an independent motel that has 50 rooms. She finds that if she charges $40 per night, all the rooms will be rented. Thereafter, for every $4 she raises the room rate, 2 fewer rooms will be rented out.

1. Make a table showing the price of the room, the number of rooms rented, and the total income for the hotel for prices of $40 to $60 per night, in $4 increments.

2. Write a function equation for the number of rooms rented as a function of the price of a room.

3. Write an equation describing the income for the motel for one night as a function of the price of a room.

4. Suppose Esperanza will be satisfied if the motel's income for a night is at least $2300. What are the possible prices she can charge to earn this income?

5. What price will earn the maximum income for the hotel? How many rooms will be rented at this price? What is the maximum income the hotel will earn?

Problem 51. Adam is raising pigs on his farm. He needs to build a rectangular pen for his pigs, and he wants to give them as much area as possible. However, he only has 180 feet of fencing.

1. Draw a diagram showing a rectangular pen, and labeling the lengh of one of the sides of the pen with a variable.

2. Write an equation that gives the area of the pen in terms of the length variable you chose.

3. How should the pen be built to get the maximum possible area? What is the maximum possible area?

Problem 52. Ernesto wants to make an open-top cardboard box to store some items on his desk. His original piece of cardboard is 9 inches by 12 inches. He needs to cut out squares from each corner of the cardboard and fold up the resulting sides to make a box, as in Figure 2.4. He wants to create the largest volume possible for his box.

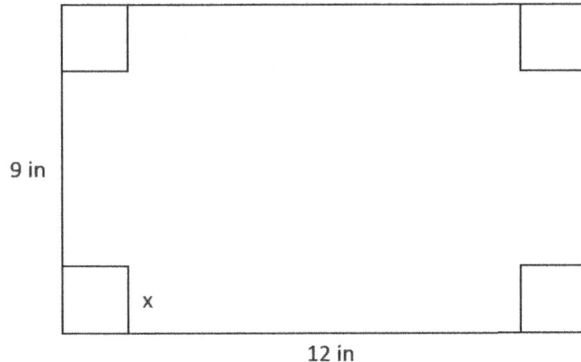

Figure 2.4: Open Top Box

1. Write an equation for the volume of the box in terms of x.

2. How large should the cutout be to maximize the volume of the box? What is the maximum volume?

Notes

2.3 Exercises

1. A movie theater seats 240 people. For any particular show, the amount of money the theater makes is a function of the number of people, n, in attendance. Analysis of recent price and attendance data suggests that for a weekday matinee showing, if the price of a ticket is set at p dollars, then the number of people in attendance, n, is given by $n = 240 - 12p$.

   (a) At what price will no one attend the showing?

   (b) How many people will attend if prices are set at the regular price of $12? What income will the theater earn at this price?

   (c) Write an equation that describes the income, I, that the theater will earn in terms of the ticket price, p.

   (d) At what price should tickets be sold to earn the greatest ticket income from the matinee show? What income will the theater earn at this price?

2. A toy rocket is launched from a table 2 feet above the ground. The height of the rocket above the ground (in feet) is given by the equation $h(t) = -16t^2 + vt + 2$, where v is the launch velocity.

   (a) If the rocket reached its maximum height of 38 feet above the ground at time $t = 1.5$ seconds, what was the launch velocity?

   (b) Write a function equation for $h(t)$.

   (c) How long was this rocket in the air?

3. A batter hits a baseball when it is 3 feet off the ground. Its height off the ground t seconds after he hits it is given by $h(t) = -16t^2 + 80t + 3$ feet. Its distance (along the ground) from home plate is $d(t) = 60t$.

   (a) Draw a graph of $h(t)$ on a domain so that the h- and t-intercepts are visible on the graph.
   (b) How long is the ball in the air?
   (c) How far from home plate does the ball hit the ground?
   (d) Use Desmos to graph the parametric equations $(d(t), h(t))$ on the domain $0 \le t \le 5.5$. Sketch the graph on your paper.

4. Adam also raises chickens, and wants to build a fence next to the chicken coop. This time, he has 100 feet of chicken wire fencing, and the fence is going to be built so that the chicken coop wall forms one side of a rectangular enclosure, with the wire fencing along the other three sides. He wants to build a rectangular space so that his chickens have as much area as possible to roam.

   (a) Draw a diagram showing the wall of the chicken coop and the three sides of wire fencing forming a rectangle, and label one of the sides of the wire fencing as x.

   (b) Write an equation for the area of the chicken enclosure in terms of x.

   (c) How should the fencing be used to get the maximum possible area? How much area will the chickens get?

5. A ball is launched vertically into the air from a height of h meters and with an initial upward velocity of v meters/second. The ball's height above ground is given by the equation $H(t) = -4.9t^2 + vt + h$, where H is in meters and t is in seconds. (This is the metric version of the gravity model.)

   (a) Write an equation to model the height of a ball thrown from a height of 2 meters off the ground, with an initial upward velocity of 40 meters/second.

   (b) How long is the ball in the air?

   (c) What is the maximum height reached by the ball?

6. Refer to the graph of k below.

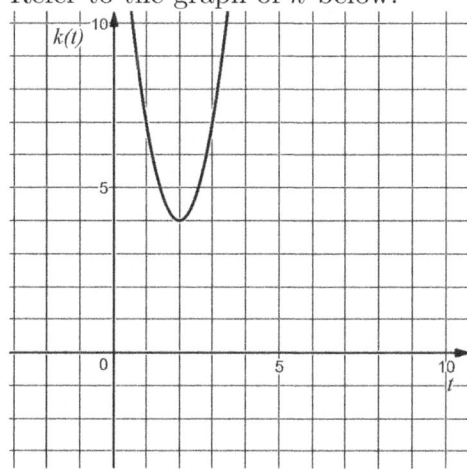

   (a) Evaluate $k(2)$.

   (b) Solve $k(t) \geq 7$.

   (c) Write an equation for $k(t)$.

   (d) Use your equation from part 6c to evaluate $k(10)$.

   (e) Use your equation from part 6c to solve $k(t) = 16$. Use Desmos to check your answer.

7. Dana is standing in a mall, with rows of stores stretching along both sides of a hall 30 meters wide. From Dana's current location in front of Foot Locker, she wants to walk to Ross, a store on the opposite side of the hall and 50 meters down the hall.

   (a) Draw a diagram showing Dana's path and impose coordinates on the diagram.

   (b) Write an equation for the line that describes the Dana's path.

   (c) Write parametric equations to describe Dana's position at time t, if it takes her 10 seconds to walk from Foot Locker to Ross.

8. Solve each equation or inequality algebraically.

   (a) $p^2 - 2p - 15 = 0$

   (b) $x^2 + 4 = 19$

   (c) $v^2 - 12v - 81 = -9$

   (d) $2x^2 - 7x - 49 > 0$

   (e) $n^2 - 3n \geq 0$

9. Let $n(x) = -x^2 + 4x - 1$

   (a) Write $n(x)$ in vertex form.

   (b) Sketch the graph of $n(x)$.

2

(c) Solve $n(x) = 0$

(d) Solve $n(x) = -1$

(e) Solve $n(x) < 2$

10. Write a linear equation to model each function below.

(a) $g(x)$ shown on the graph below.

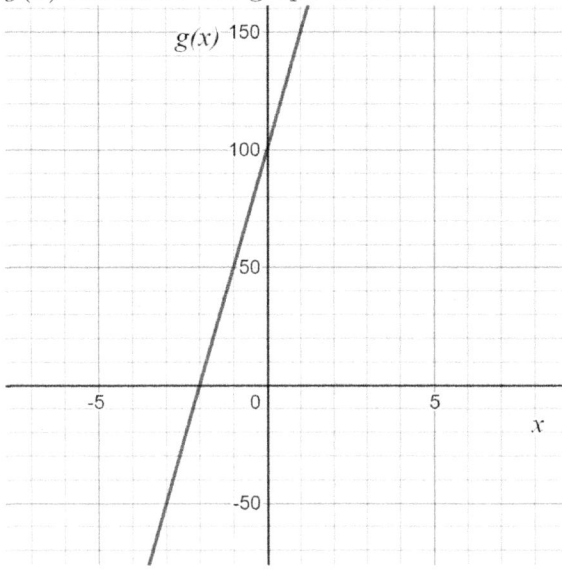

(b) $n(x)$ where $(-2, 1)$ and $(4, -2)$ lie on the graph of $n(x)$.

(c) $q(x)$ defined by the table below.

x	-1	0	1	2	3
$q(x)$	$-\frac{7}{3}$	-2	$-\frac{5}{3}$	$-\frac{4}{3}$	-1

Chapter 3

Exponential and Logarithmic Functions

3.1 Exponential Functions

Goals:

- E: Be able to solve an equation with an unknown exponent.

- E: Be able to determine the equation of an exponential function given a table of values.

- F: Be able to give the solution to an inequality or set of inequalities using proper mathematical notation.

Definition 53. An **exponential function** is a function of the form $f(t) = ab^t$, where a is a nonzero real number, and b is a positive real number not equal to 1. Whenever we have an expression like b^t, b is called the **base** and t is the **exponent**. The number e is often used as the base in exponential functions because it has a number of special properties (you will learn more about this in calculus). The value of e is *about* 2.718.

Note: In an exponential function, the variable is in the exponent. For example $f(x) = 2^x$ is an exponential function, but $g(x) = x^2$ is not an exponential function because the exponent is not a variable.

Investigation 54. Two companies that rent laptops have different late fee policies.

Company 1: For each day the laptop is late, you owe an additional \$5. On day 1, your total late penalty is \$5. On day 2, your total late penalty is \$10. On day 3, your total late penalty is \$15, and so on.
Company 2: For each day the laptop is late, your penalty doubles from the previous day. On day 1, your late penalty starts at \$0.25. On day 2, your late penalty doubles to \$0.50. On day 3, the late penalty doubles to \$1, and so on.

As a customer, which company do you think has the better late fee policy? Explain your reasoning.

Investigation 55. Let $f(x) = a \cdot b^x$.

In Desmos, graph f and create sliders for a, and b. What role do each of these constants play? Include sketches and verbal descriptions to help explain the role of each constant.

Example 56. $f(x) = 2^x$ is an exponential function. A table of values for $f(x)$ is shown in Table 3.1. Recall that a negative exponent means taking the reciprocal of the positive exponent, so $f(-1) = 2^{-1} = \frac{1}{2^1} = \frac{1}{2}$. Graph f on the domain $-1 \le x \le 5$. Describe the shape of the graph.

x	$f(x)$
-1	$\frac{1}{2}$
0	1
1	2
2	4

Table 3.1: Values of f

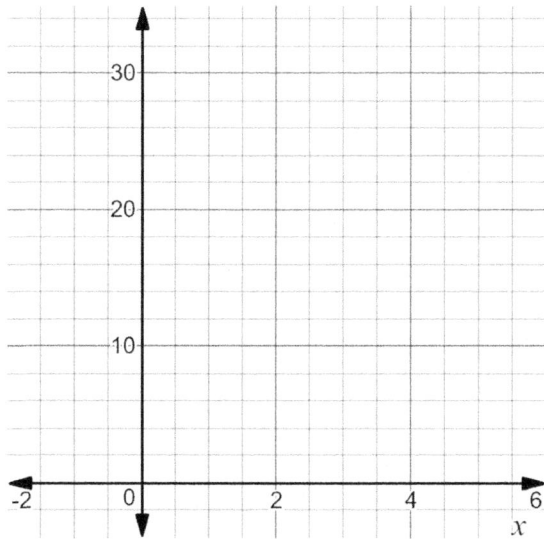

Problem 57. Refer to $P(t)$ as given in Table 3.2.

t	0	2	4	6
$P(t)$	10	90	810	7290

Table 3.2: Values of P

1. Assume that $P(t)$ is an exponential function, and write a function equation for $P(t)$.

2. What is $P(3)$?

3. Solve $P(t) \leq 10,000$.

Problem 58. Refer to $F(x)$ as shown in Table 3.3.

x	-4	-2	0	2
$F(x)$	80	20	5	1.25

Table 3.3: Values of F

1. Write an exponential equation for F.

2. Evaluate $F(5)$.

3. Solve $F(x) = 2000$.

Problem 59. Let $g(x) = 10 \cdot 2^x$, $h(x) = 20^x$, $k(x) = 10^x 2^x$, and $m(x) = (10 \cdot 2)^x$. By graphing these functions, decide which, if any, of these are really the same function.

Problem 60. Consider $2^x 3^y$, 6^{xy}, and 6^{x+y}. By plugging in pairs of values for x and y, decide whether any of these are the same.

Notes

3.1 Exercises

1. Refer to $P(t)$ as given in Table 3.4.

t	0	3	6	9
$P(t)$	5	135	3645	98415

Table 3.4: Values of $P(t)$

    (a) Assume that P is an exponential function, and write a function equation for $P(t)$.

    (b) What is $P(4)$?

    (c) Solve $P(t) \geq 2,000$.

2. Refer to $g(x)$ as given in Table 3.5.

x	1	2	3	4
$g(x)$	0.8	0.16	0.032	0.0064

Table 3.5: Values of $g(x)$

    (a) Assume that $g(x)$ is an exponential function, and write a function equation for $g(x)$.

    (b) What is $g(7)$?

    (c) Solve $g(x) \geq 0.01$.

3. Suppose you put \$100 into a savings account paying 2.5% interest each year (compounded annually).

    (a) What percent will you earn if you leave the money in the account for 3 years?

    (b) Explain why the answer is NOT $2.5\% \times 3 = 7.5\%$.

    (c) One student solved the problem this way: $100 \times 1.025 \times 1.025 \times 1.025 = 107.69$. $107.69 - 100 = 7.69$. So the answer is 7.69%. Explain this student's work. What does 7.69% represent in this problem?

4. The population of mosquitoes on a small island increases during the wet season. The population was measured once per week, as shown in Table 3.6.

t	0	1	2	3
$S(t)$	100	1600	25600	409600

Table 3.6: Mosquito population, $S(t)$

    (a) What is the average rate of growth of the mosquito population over the three weeks shown in the table?

    (b) Assuming that the population continues to grow exponentially, write a function equation for $S(t)$.

(c) When will the population reach 1 million (1,000,000)?

5. Maria is preparing envelopes for mailing. Maria has already prepared 50 envelopes this morning. At 1 pm, she returns from lunch. She can prepare 110 envelopes per hour.

    (a) Write an equation for the total number of envelopes Maria has prepared t hours after 1 pm.

    (b) If Maria hopes to prepare 750 envelopes before going home, when can she expect to be done?

6. Recall that the vertical position of falling objects can be modeled by the equation $h(t) = -16t^2 + vt + c$, where t is the time in seconds, $h(t)$ is the height in feet, c is the initial height of the object, and v is the initial velocity of the object. A football is kicked from a height of 3 feet above the ground and has an initial velocity of 60 feet per second. [Note: Footballs are affected by air resistance (wind, etc), so this is not a very accurate model. There are more complicated models that take these factors into account.]

    (a) What is the highest the football will go? At what time does this happen?

    (b) How long is the football in the air?

7. The graph below shows the height above ground, h (in meters), of a Ferris wheel rider t seconds after her ride starts (when she is at the 6 o'clock position on the wheel).

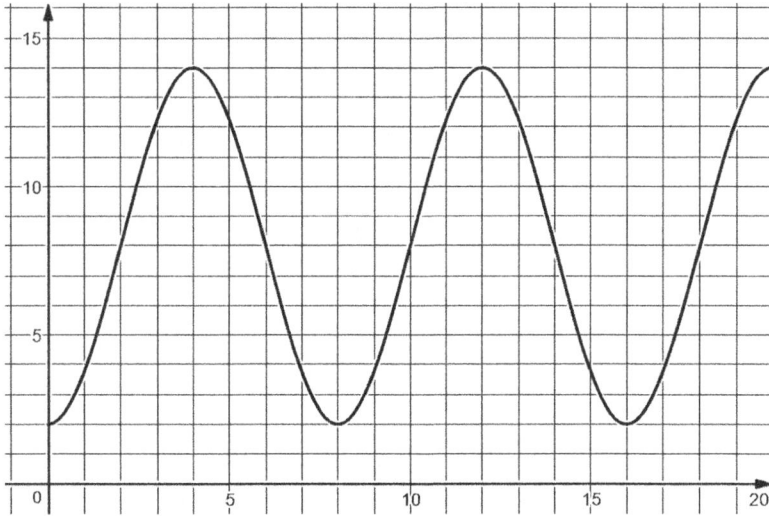

    (a) How far above the ground is the rider at $t = 6$ seconds?

    (b) What are the maximum and minimum heights reached by the rider? What do these heights tell you about the radius of the Ferris wheel?

    (c) When is the rider 6 meters above ground?

    (d) When will the rider be less than 8 meters above ground?

    (e) Based on the graph of h, how long does it take the rider to make one complete trip around the wheel?

3.2 Exponential Modeling

Exponential functions are used to model situations in which the rate of growth of an quantity is proportional to the quantity's size. Exponential models are used in situations such as money earning compound interest, population growth, and radioactive decay.

Goals:

- E: Be able to solve an equation/inequality with an unknown exponent.

- E: Be able to model a situation with appropriate exponential equation(s) and interpret the solution.

- E: Be able to determine the equation of an exponential function given a table of values.

- F: Be able to compute the average rate of change of a given function on a given interval.

Example 61. Pam opens a banking account with $500. The account earns 1.5% compounded annually. Let t be the number of years the bank account has been open, and $B(t)$ the balance in the account.

1. Make a table showing the balance, $B(t)$, in the account at $t = 0$, 1, 2, 3, and 4 years.

2. Write a formula for the function B.

3. Use your formula for B to determine the balance in 20 years.

4. When will the balance in the account reach $800?

Problem 62. Tyus opens a banking account with $800. The account earns 3% compounded annually. Let t be the number of years the bank account has been open, and $A(t)$ the balance in the account.

1. Make a table showing the balance, $A(t)$, in the account at $t = 0$, 1, 2, 3, and 4 years.

2. Write a formula for the function A.

3. Use your formula for the function A to determine the balance in 20 years.

4. When will the balance in the account reach \$1400?

Problem 63. The bacteria in a dish have an initial population of 1000, and are growing such that the population doubles every 45 minutes. Let t be the number of minutes that have passed since the initial population was measured, and let $P(t)$ be the population at time t minutes.

1. Make a table showing the population, $P(t)$, at times $t = 0$, 45, 90, and 135 minutes.

2. Write an equation for the function P.

3. Use your function equation to determine the population of bacteria in 6 hours (360 minutes).

4. When will there be 250,000 bacteria?

5. Determine the average number of bacteria added per hour in the first 6 hours.

Problem 64. A Christmas tree lot sold 200 trees in 2010 and 350 trees in 2015.

1. Assuming the number of trees sold is increasing exponentially, write a formula for the number of trees sold t years after 2010.

2. If the number of trees sold continues to increase exponentially, how many trees will be sold in 2019?

3. If the number of trees sold continues to increase exponentially, in what year will 500 trees be sold?

Problem 65. The population of Dry Gulch has been decreasing by $\frac{1}{2}$ every 20 years. In 1910 the population of Dry Gulch was 3800 people.

1. Write a formula for the population of Dry Gulch t years after 1910.

2. If the number of people in Dry Gulch continued to decrease exponentially, how many people were there in the town in 2010?

3. If the number of people in Dry Gulch continues to decrease exponentially, when will the town become a ghost town (a population of less than 1 person)?

Problem 66. A camera costs \$110 now. The cost of the camera increases by 6% annually.

1. Write a formula for the cost of the camera t years from now.

2. If the cost continues to rise exponentially, how much will the camera cost in 3 years?

3. If the cost continues to rise exponentially, how long will it take the cost of the camera to reach $200?

Problem 67. Carbon dating is used to determine the age of bones, tools and other relics. Carbon-14 has a half life of 5728 years, meaning that if an object is found to have 50% of its original carbon, it is 5728 years old.

1. Make a table for the amount of Carbon-14, $C(t)$, at time $t = 0$, 5728, and 11456 years, supposing that the initial amount of carbon is an unknown a.

2. Write an equation for the amount of Carbon-14, $C(t)$, at time t.

3. In 1990 a body was found in the Sierra Nevada mountain range. An examination of the tissue found that 27% of the carbon-14 present at the time of death had decayed. How long ago did the man die?

Notes

3.2 Exercises

1. The population of a small town has been growing. The population is shown in Table 3.7.

t (years since 1980)	0	10	20	30
$P(t)$	1000	1344	1806	2427

Table 3.7: Population, $P(t)$

   (a) What class of functions (linear, quadratic or exponential) would be best to model the population of the town. Be sure to explain your reasoning.
   (b) What is the average number of people added to the town per year between 1980 and 1990?
   (c) Assuming that the population continues to grow exponentially, write a function equation for $P(t)$.
   (d) If the population continues to grow according to the model, what is the population in 2015?
   (e) When will the population reach 5,000?

2. The pesticide DDT was used in the US and later banned. The half-life of DDT is about 15 years.

   (a) Write an exponential model for the amount of DDT, $A(t)$, remaining after t years, if the initial sample is 100 grams.
   (b) According to your model, how much of the initial sample will remain after 60 years?
   (c) How many years will it take for the sample to decay to 1 gram?

3. Plutonium 238 is a radioactive element that decays at a rate of 0.8% per year.

   (a) What percentage of an initial supply of 500 grams Plutonium 238 will remain after 40 years?
   (b) How many years will it be until an initial supply of 500 grams of Plutonium 238 has decayed to half of its initial mass?

4. When buying a new car, one consideration is how fast the car loses value, known as the depreciation rate. For example, one version of the Jeep Liberty loses value more rapidly than some of its competitors. From the initial purchase price of $23,395, an owner can expect the value of the car 5 years later to be $15,239.

   (a) What is the average dollar value decline per year during the first 5 years of ownership?
   (b) Use an exponential model produce a function that gives the value of the Jeep in terms of the number of years, t, since the Jeep was new.
   (c) Use your model to predict the value of the Jeep when it was 3 years old.
   (d) When should the value of the Jeep decline to $5,000?

5. A river where salmon spawn had 2214 salmon spawn in 2015. In 2018, only 2023 salmon spawned in the same river.

   (a) Assuming the number of salmon spawning is decreasing exponentially, write a formula for the number of salmon spawning in the river t years after 2015.
   (b) If the number of salmon spawning in the river continues to decrease exponentially, how many salmon will spawn in the river in 2025?

(c) If the number of salmon spawning in the river continues to decrease exponentially, in what year will the number of salmon spawning in the river decrease to half of the number of 2015?

6. Oscar charges $5 per linear foot to paint any standard outdoor wooden fence, plus $20 to cover incidental items, such as brushes.

    (a) Write a function equation that gives the cost to paint a fence of length L feet.

    (b) What is the cost to paint a fence 50 feet long?

    (c) If Oscar recently painted a fence and charged $110, how long was the fence?

7. For a certain species of shark, its length (in feet) varies according to the equation

$$l(w) = kw^{\frac{1}{3}}$$

where w is the shark's weight in pounds.

    (a) If a 6 foot shark weighs 200 pounds, write an equation for $l(w)$.

    (b) Using your equation from part 7a, how long will a shark be if it weighs 400 pounds?

    (c) Using your equation from part 7a, how much will a shark weigh if it is 4 feet long?

8. Let $f(x) = -x^2 + 4x$ and $g(x) = x - 4$.

    (a) Evaluate $f(-2)$

    (b) Solve $g(x) < -4$

    (c) Solve $f(x) = 0$

    (d) Solve $f(x) \leq 0$

    (e) Solve $f(x) = g(x)$

    (f) Solve $g(x) < f(x)$

    (g) Solve $f(x) = 1$

3.3 Comparing Linear and Exponential Models

Both linear and exponential functions are often used to model real data. In this section, we will explore how each of these models works in the context of population data.

Goals:

- E: Be able to model a situation with appropriate exponential equation(s) and interpret the solution.

- E: Be able to determine the equation of an exponential function given a table of values.

- F: Be able to determine an appropriate function class (linear, quadratic, exponential, trigonometric) to model a particular situation.

- F: Be able to compute the average rate of change of a given function on a given interval

Definition 68. The **average rate of change** of a function h defined on an interval $a \leq x \leq b$ is

$$\frac{h(b) - h(a)}{b - a}.$$

Note that the average rate of change computes the slope of the line between the two points $(a, h(a))$ and $(b, h(b))$.

Problem 69. In Exercise Set 1, Problem 1, we saw that the height above the ground, h of a ball t seconds after being dropped off a 60 meter tall building is given by the equation $h(t) = -4.9t^2 + 60$ and is shown on the graph below.

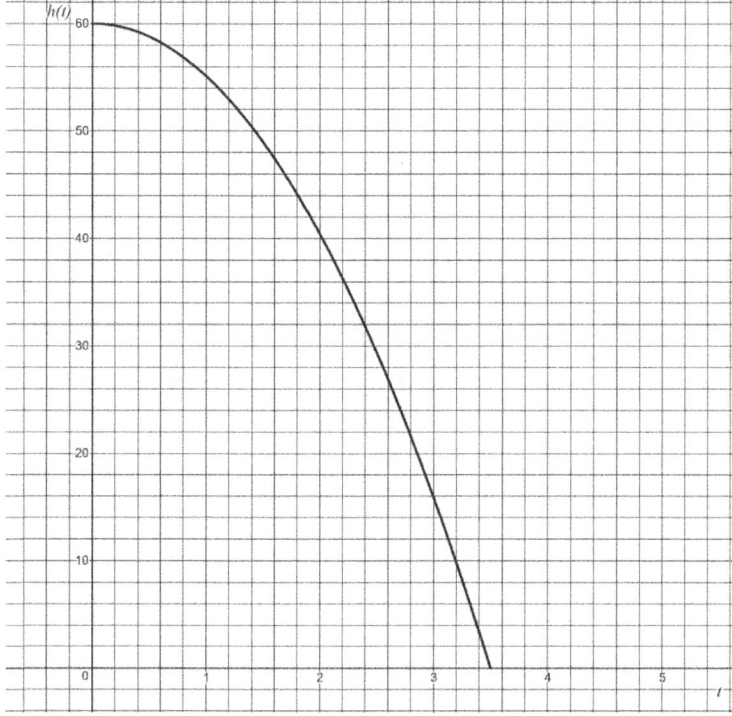

1. Between when the ball is dropped and when it hits the ground, does it fall at a constant rate? Why or why not?

2. How far does the ball fall between $t = 0$ and $t = 3$ seconds?

3. What is the average rate of change of the ball between $t = 0$ and $t = 3$ seconds? Write an algebraic expression using function notation for the average rate of change, then compute it.

4. What are the units of the change in height of the ball? What are the units in the change in time? What are the units for the average rate of change you computed in part 3?

5. What is the average rate of change of the ball between the time it is dropped off the building and the time it hits the ground? Be sure to include units in your answer.

Definition 70. A function h is **increasing** on the interval $a \leq x \leq b$ if, whenever two points c and d are in the interval, with $c < d$, then $f(c) < f(d)$. A function h is **decreasing** on the interval $a \leq x \leq b$ if, whenever two points c and d are in the interval, with $c < d$, then $f(c) > f(d)$.

Informally, the definitions for increasing and decreasing are stating that if a function goes up as we move from left to right, it is increasing, and if it goes down as we move from left to right, it is decreasing.

Below, sketch a function that is increasing on $(-\infty, \infty)$ and a function that is decreasing on $(-\infty, \infty)$.

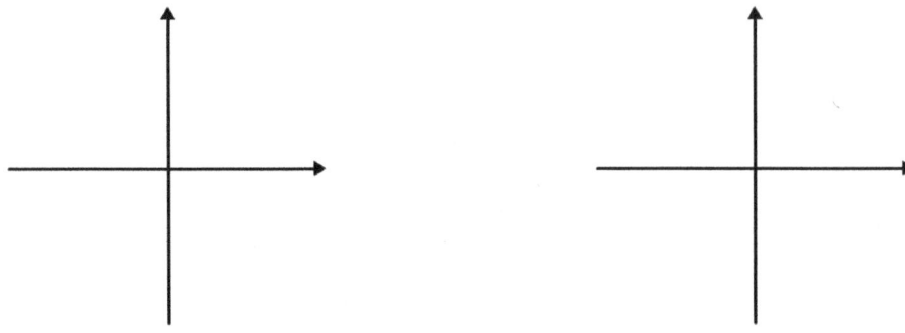

Example 71. Refer to the graph and table for g in Figure 3.1.

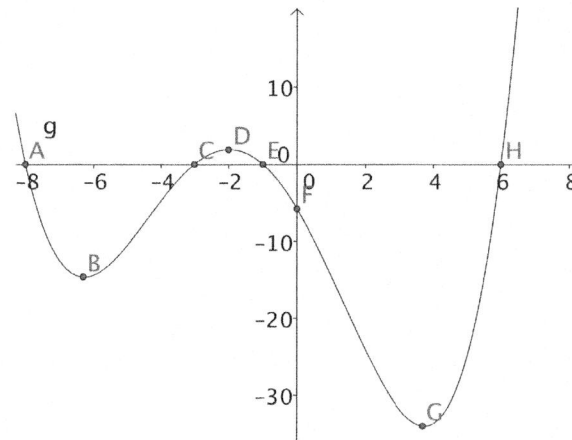

	A	B	C	D	E	F	G	H
x	-8	-6.3	-3	-2	-1	0	3.7	6
$g(x)$	0	-14.6	0	1.9	0	-5.8	-33.9	0

Figure 3.1: Graph and table for g

1. Find the average rate of change of g on the interval $-6.3 \le x \le 3.7$.

2. Find the average rate of change of f on the interval $-8 \le x \le 6$.

3. On what intervals is g increasing? On what intervals is g decreasing?

The solution to Example 71 is outlined below:

The function we are given goes down (is decreasing) from point A to point B, then increases from point B to point D, decreases again from point D to point G, and then increases from point G to point H and beyond. In this case, we only have the graph, and not the equation, so we should not make assumptions about what happens outside of the interval $-8 \le x \le 6$. Thus, g is increasing on the intervals $-6.3 \le x \le -2$, and again on $3.7 \le x \le 6$. The function g is decreasing on the intervals $-8 \le x \le -6.3$ and $-2 \le x \le 3.7$. The average rate of change of g on the interval $-6.3 \le x \le 3.7$ is $\frac{-33.9-(-14.6)}{3.7-(-6.3)} = \frac{-19.3}{10} = 1.93$. The average rate of change of g on the interval $-8 \le x \le 6$ is $\frac{0-0}{6-(-8)} = 0$. Can you explain why the average rate of change was 0 on the interval $-8 \le x \le 6$?

Problem 72.

1. Using the graph of $f(x) = \frac{1}{3}x + 1$ below, choose two points on the graph and find the average rate of change between those two points.

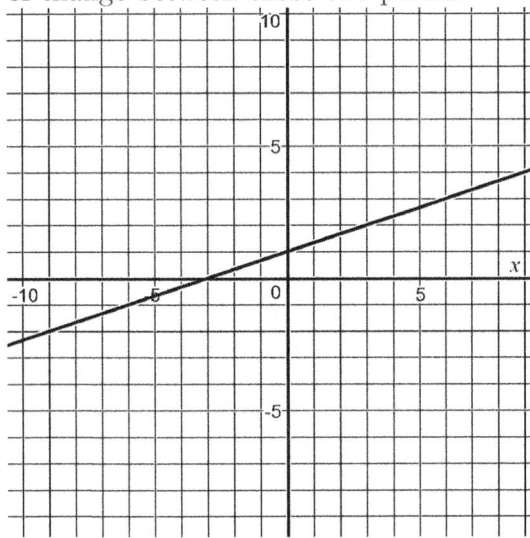

2. Choose two different points on the graph of $f(x) = \frac{1}{3}x + 1$ above and find the average rate of change between those two points.

3. Using the graph of $g(x) = 2^x$ below, choose two points on the graph and find the average rate of change between those two points.

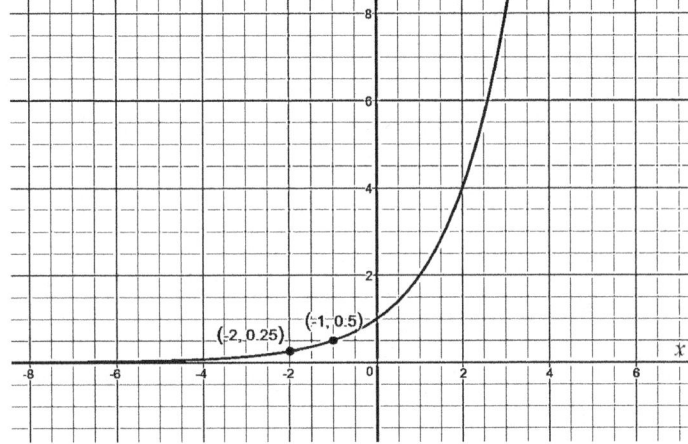

4. Choose two different points on the graph of $g(x) = 2^x$ above and find the average rate of change between those two points.

5. Summarize what you have learned about how linear functions like f from part 1 and exponential functions like g from part 3 are different in terms of their rates of change.

Investigation 73. You have been hired as a consultant for the City of Los Angeles. City planners need to understand population data and use the data to predict the population in the future. Having a good estimate of the city's population helps the planners decide how much money will be needed for city maintenance and services, and helps local officials anticipate the needs for housing, school and other construction, and much more. Your work begins with finding the population of Los Angeles in the United States Census for 1920 to 2010. This data appears in Table 3.8.

x (yrs since 1920)	Year	Actual Pop.	Pop. Using Linear Model	Pop. Using Exponential Model
0	1920	576,673		
10	1930	1,238,048		
20	1940	1,504,277		
30	1950	1,970,358		
40	1960	2,479,015		
50	1970	2,816,061		
60	1980	2,966,850		
70	1990	3,485,398		
80	2000	3,694,820		
90	2010	3,792, 621		

Table 3.8: U.S. Census Bureau data for Los Angeles (city), 1920-2010

Problem 74. Linear Model. One way to model a population is with a linear function, $y = mx + b$, where m and b are real numbers. As a first attempt, find the equation of the line passing through the data for 1920 and 2010.

Problem 75. Use your linear model from Problem 74 to put the estimated population for values $t = 0$ to $t = 90$ into Table 3.8 . How do the values compare with the actual data?

Problem 76. At this url: `https://goo.gl/dynoa2`, you will find Desmos set up with the data. Using Desmos, find the values of m and b that you think make the line $y = mx + b$ best fit the data. Make a note of the equation. How does the equation compare to your linear model from Problem 74? How does the equation look compared to the actual data points on the graph?

Problem 77. Exponential Model. Find the equation of an exponential model passing through the data for 1920 $(t = 0)$ and 2010 $(t = 90)$.

Problem 78. Use your exponential model from Problem 77 to add the estimated population for values $t = 0$ to $t = 90$ to Table 3.8 . How do the values compare with the actual data?

Problem 79. Graph the exponential model $G(x) = 945802(1.01845)^x$ in Desmos. How does the equation compare to your own linear model from Problem 74? How does the best-fit equation look compared to the actual data points on the graph?

Problem 80. Comparing the models. Your final job is to make a decision about which model will best predict the population over the next ten to twenty years. Justify your decision based on both mathematics and what you know or think will happen in Los Angeles.

Problem 81. For each table or situation below, identify which class of function (linear, quadratic or exponential) you would choose to model that table or situation, and explain why. Then write a function equation for each table or situation.

1. Let $t(n)$ be a function that gives the number of squares in the nth figure for the pattern shown below.

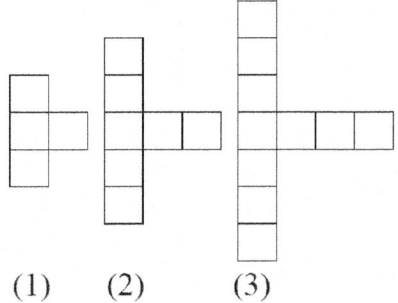

(1)　　(2)　　　(3)

2.

x	-4	-2	0	2	4
$g(x)$	48	12	3	0.75	0.1875

3. A breeder keeps pet rabbits. In January he has 20 rabbits. His population of rabbits doubles every 3 months. Let P, the population of rabbits, be a function of t, the number of months since January.

4.

x	-6	-3	0	3	6
$f(x)$	600	100	-400	-900	-1400

5. Let $k(n)$ be a function that gives the number of squares in the nth figure for the pattern shown below.

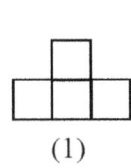   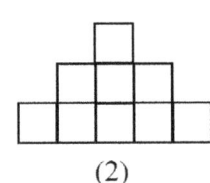   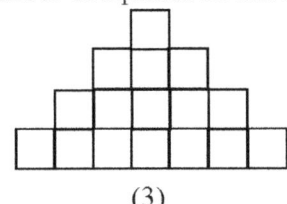

(1)   (2)   (3)

6. Chris has \$150 on his meal card. Every time he orders a meal, \$9 is deducted from his card. Let M, the balance Chris has on his meal card, be a function of n, the number of meals he has ordered.

Notes

3.3 Exercises

1. Refer to the population of Las Vegas, NV, for given years as shown in Table 3.9.

t	0	10	20	30
$P(t)$	164,674	259,834	484,487	583,756

Table 3.9: Population, $P(t)$, of Las Vegas t years after 1980

  (a) Using the population of Las Vegas in 1980 and 2010, build an exponential model for $P(t)$.

  (b) Use your exponential model to predict the population of Las Vegas in 1990. How does your prediction compare with the actual population at that time?

  (c) Use your exponential model to predict in what year the population of Las Vegas will be 600,000.

  (d) Using the population of Las Vegas in 1980 and 2010, build a linear model for $P(t)$.

  (e) Do you think the linear model or the exponential model is a better fit to the data? Explain.

  (f) What could happen in the future that would make the model fail?

2. Let m be defined as shown in the graph below.

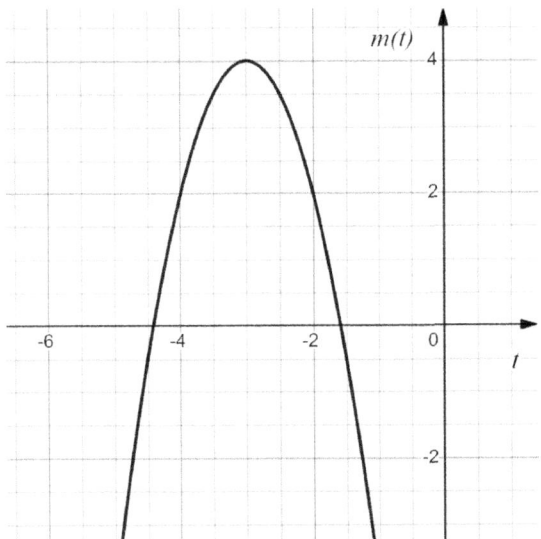

  (a) What is the average rate of change of m on the interval $[-4, -2]$?

  (b) What is the average rate of change of m on the interval $[-3, -2]$?

  (c) On what interval(s) is m increasing?

  (d) On what interval(s) is m decreasing?

  (e) What are the domain and range of m?

3. Let $k(x) = x^2 - 3$.

  (a) What is the average rate of change of k on the interval $[0, 2]$?

  (b) What is the average rate of change of k on the interval $[-6, 4]$?

(c) On what interval(s) is k increasing?

(d) On what interval(s) is k decreasing?

(e) What are the domain and range of k?

4. The town of Allen had a population of 36,000 in 1980, and has been growing at 2.2% per year. The town of Berry had a population of 44,200 in 1980, and a population of 48,000 in 1990.

(a) Find an exponential model, $A(t)$, for the population of Allen t years after 1980.

(b) Find an exponential model, $B(t)$, for the population of Berry t years after 1980.

(c) When did the population of Allen reach 45,000?

(d) When did the population of Allen equal the population of Berry?

(e) What is the average number of people added to Allen between 1980 and 2000?

5. The owners of a movie theater know that the number of people who attend is a function of the price of the tickets. The theater has a capacity of 240 people. If tickets are sold for $1, the theater will fill up completely. On the other hand, if the theater charges $21/ticket, no one will buy tickets.

(a) Assuming that the number of tickets sold, $S(t)$ is a linear function of the price, t, write an equation for $S(t)$.

(b) At what price will the theater fill 150 seats?

6. A baseball is launched into the air from a height of 2 meters and with an initial upward velocity of 25 meters/second. The ball's height above ground is given by the equation $H(t) = -4.9t^2 + vt + h$, where H is in meters and t is in seconds.

(a) Write an equation to model the height of a ball.

(b) How long is the ball in the air?

(c) What is the maximum height reached by the ball?

(d) On what interval is the height of the ball increasing?

3.4   Combining Functions

Often in mathematics, we are interested in using existing functions to create new functions. In this section and the next, we will look at some common ways of building new functions.
Goals:

- F: Be able to determine a composition of functions given in any form (graph, table, equation).

- F: Be able to perform arithmetic (sum, difference, product, quotient) on functions given in any form (graph, table, equation).

- F: Be able to identify functions and operations that could be combined to produce a given function.

Example 82. You have already seen that function notation uses parentheses to mean something other than multiplication. When examining functions, the function use of parentheses will be the norm. For instance, if $f(x) = 2x$ and $g(x) = 3x^2 - 2x - 5$, then we can define $h(x) = f(x) + g(x)$.

1. Find $h(6) = f(6) + g(6)$.

2. Find an explicit formula for $h(x)$ (a formula for h that does not use f and g). Use this equation to find $h(6)$. How does this compare to your answer in part 1?

Problem 83. Refer to f and g from Example 82. Define another new function $k(x) = 2f(x) - 4g(x)$. Evaluate $k(6)$. Then find an explicit formula for $k(x)$ (a formula for k that does not use f and g). Use your formula to evaluate $k(6)$.

Problem 84. Refer to f and g from Example 82. Define $m(t) = \frac{f(t)}{g(t)}$. Evaluate $m(6)$. Then find an explicit formula for $m(t)$ (a formula for m that does not use f and g).

Another way to build a new functions is by using the output of one function as the input of another. This is called function composition.

Definition 85. Given functions f and g, define the **composition** $p(x) = (f \circ g)(x) = f(g(x))$.

Example 86. Refer to f and g from Example 82. Define $p(x) = f(g(x))$.

1. Find $p(6) = f(g(6))$.

2. Find an algebraic formula for $p(x)$. Use this formula to find $p(6)$.

3. Define $q(x) = g(f(x))$. Find $q(6) = g(f(6))$. Find an algebraic formula for $q(x)$. Compare these to your answers in parts 1 and 2. In this case, does it appear that $(f \circ g)(x) = (g \circ f)(x)$?

Problem 87. Answer the following questions using the tables for $a(x)$, $b(x)$ and $c(x)$ given below.

x	-2	-1	0	1	2	3
$a(x)$	-1	1	2	5	-4	-1

x	1	2	5	6
$b(x)$	2	3	-2	1

t	-2	0	1	3
$c(x)$	5	0	-1	2

1. Evaluate $a(b(1))$.

2. Evaluate $b(a(1))$.

3. Evaluate $a(b(c(3)))$.

4. Let $h(x) = a(x) \cdot b(c(x))$. Evaluate $h(-2)$.

5. What is x if $a(b(x)) = 5$?

6. What is x if $c(b(x)) = 2$?

Problem 88. For this problem, refer to $a(t) = 4t - 7$, $b(t) = -2 \cdot 5^t$, and $r(t)$ as given in Table 3.10.

t	0	1	2	3
$r(t)$	25	60	1	-10

Table 3.10: $r(t)$

1. Let $c(x) = \frac{2b(x)-3}{a(x)}$. Evaluate $c(3)$. Then find an algebraic formula for $c(x)$.

2. Let $d(x) = x \cdot a(b(x))$. Evaluate $d(2)$. Then find an algebraic formula for $d(x)$.

3. Let $s(t) = a(r(t))$. Evaluate $s(3)$.

Problem 89. For each function below, determine functions f and g so that the function can be written as an operation (sum, difference, product, etc) or composition of f and g.

1. $k(x) = e^{1/x^2}$
2. $a(x) = \frac{1}{x^2+2}$
3. $b(x) = \frac{x^2}{10^x}$
4. $c(x) = x^2(x - 2x^2)$

5. $d(x) = 2^x - \frac{1}{2}x^3$
6. $m(x) = \sqrt[3]{x^4 + 8}$
7. $n(x) = x^2 e^{x^2}$

Notes

3.4 Exercises

1. Let $q(x) = x - 2$. Use the graph of $k(x)$ and the table of $p(x)$ shown below and the equation of $q(x)$ to answer the following questions.

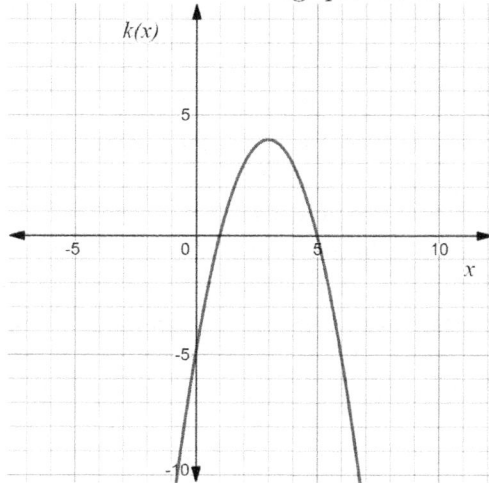

x	-4	-2	0	2	4
$p(x)$	1	2	3	4	5

(a) Let $f(x) = (k \circ p)(x)$ Evaluate $f(-2)$.

(b) Solve $f(x) = 4$ for x.

(c) Let $g(x) = q(k(x))$. Evaluate $g(0)$.

(d) Solve $g(x) = 1$ for x.

(e) Let $a(x) = p(x) - 2q(x)$. Evaluate $a(-4)$.

(f) Let $b(x) = \frac{k(x)}{(q(x))^2}$. Evaluate $b(6)$.

(g) What is the average rate of change of p on the interval $[-4, 2]$?

2. Let $v(t) = 3t - 4$, $p(t) = e^{3t}$, $n(t) = t^2 - 1$. Find an algebraic formula for each function below.

(a) $c(t) = v(p(t))$

(b) $h(t) = p(v(t))$

(c) $d(t) = n(v(t))$

(d) $k(t) = v(n(t))$

(e) $e(t) = -v(t) + 2(n(t))$

(f) $f(t) = \frac{n(t)}{v(t)}$

(g) $g(t) = v(t) \cdot n(t)$

3. For each function below, determine functions f and g so that the function can be written as an operation (sum, difference, product, etc) or composition of f and g.

(a) $r(x) = (3x + 1)^2$

(b) $s(x) = 2\sqrt{x} - \frac{1}{\sqrt{x}}$

(c) $q(x) = \frac{e^x}{x^2-1}$

(d) $w(x) = x^2 \cdot e^{x^2}$

(e) $y(x) = x^3 \cdot e^{x^2}$

4. Cell phones begin to lose (resale) value immediately after purchase. After one year, an iPhone can be resold at 63% of its list price. Assume that each year after that, the resale value continues to drop to 63% of the resale price from the previous year.

   (a) Assuming the list price of an iPhone is $649, write an exponential model for the resale price, $P(t)$, of an iPhone t *months* after purchase.
   (b) According to your model, what is the resale value of the iPhone after 18 months?
   (c) According to your model, when is the resale value of the iPhone $150?
   (d) Assume that a particular Android phone has a value, $A(t)$, that is always $100 less than the iPhone. Write $A(t)$ as a composition of $P(t)$ and another function, $h(t)$, so that $A(t) = h(P(t))$.

5. Consider the functions $k(x) = 8^x$ and $m(x) = 4 \cdot 2^x$.

   (a) Are $k(x)$ and $m(x)$ the same function? Explain.
   (b) Write $m(x)$ as a composition of $f(x) = 2^x$ and another function $b(x)$, so that $m(x) = b(f(x))$.
   (c) Write $k(x)$ as a composition of $f(x) = 2^x$ and another function $d(x)$, so that $k(x) = d(f(x))$.

6. A new sofa cost $1200 in 2010. The value of the couch decreases by 15% per year.

   (a) Write a formula for the value of the couch t years after 2010.

   (b) If the cost continues to decrease exponentially, what will the value of the sofa be in 5 years?

   (c) If the cost continues to decrease exponentially, how long will it take for the couch to reach half of its original price.

   (d) How much does the value of the couch decrease per year in the first five years?

7. Adam, the farmer, is building a third rectangular enclosure, this one for his cows and bulls. He has 600 feet of fencing. He wants to enclose all four walls with fence, but he also wants to build another wall of fencing through the middle of the enclosure to separate the cows from the bulls. As before, he wants to use his fencing to give the animals as much total area as possible.

   (a) Draw a diagram showing the enclosure with the dividing fence, and label one of the sides of the wire fencing as x.

   (b) Write an equation for the area of the enclosure in terms of x.

   (c) How should the fencing be used to get the maximum possible area? Assuming each gets half of the space, how much area will the cows and the bulls each get?

8. Let $f(x) = -2x - 5$ and $g(x) = x^2 + 6x + 2$. Solve each inequality below algebraically.

   (a) $f(x) < 6$
   (b) $g(x) \geq 2$
   (c) $f(x) > g(x)$

3.5   Exponents & Radicals

Goals:

- E: Be able to apply exponents and radicals to simplify expressions.
- F: Be able to identify functions and operations that could be combined to produce a given function.
- F: Be able to determine a composition of functions given in any form (graph, table or equation).
- F: Be able to perform arithmetic operations on functions given in any form.

Definition 90. Below is a list of exponent rules where $a \neq 0$, $b \neq 0$, and m and n are real numbers. For each rule, write an argument that justifies the rule.

1. $a^n \cdot a^m = a^{m+n}$

2. $\frac{a^m}{a^n} = a^{m-n}$

3. $(a^m)^n = a^{mn}$

4. $(ab)^m = a^m \cdot b^m$

5. $a^{-n} = \frac{1}{a^n}$

6. $a^0 = 1$

Problem 91. Use the exponent rules from above to simplify the following expressions. Write your answer without negative exponents.

1. $\frac{x^2 y^3}{xy^5}$

2. $\frac{1}{t^{-5}}$

3. $5x^2 y (2x^4 y^{-3})^2$

4. $\left(\frac{-3a^2 bc^{-2}}{-a^3 b^2 c^2}\right)^3$

5. $\left(\frac{8x^2 y^{-2}}{3z^{-4}}\right)\left(\frac{-9z^{-2}}{4x^3 y^{-1}}\right)$

6. $(3x^a y^b z^c)(-y^f z^g)$

Example 92. Let $a(x) = 10^x$, $b(x) = 10^{2x^2}$, $c(x) = x^3$. Write a formula for each function below, using the exponent rules to simplify the formula and write without negative exponents.

1. $k(x) = (c \circ b)(x)$

2. $f(x) = (c(x) \cdot a(x))^4$

3. $g(x) = (b(x))^{-2}$

4. $h(x) = 2\left(\frac{a(x)}{b(x)}\right)$

5. $j(x) = a(x) \cdot b(x)$

Problem 93. Let $w(x) = 3e^x$, $y(x) = e^{-x}$, $z(x) = \frac{1}{x^2}$. First, write the given function as a combination of w, y, and z, using any arithmetic operations and the operation of composition, as appropriate. Then, use the exponent rules to simplify the formula if possible. For example,

$$f(x) = 3e^{e^{-x}}$$

is the composition of $w(x)$ and $y(x)$, so

$$f(x) = (w \circ y)(x) = w(y(x))$$

1. $m(x) = e^{-x} \cdot 3e^x$

2. $h(x) = \frac{3e^x}{e^{-x}}$

3. $g(x) = \frac{1}{(e^{-x})^2}$

4. $k(x) = \left(\frac{3e^x}{x^2}\right)^2$

5. $j(x) = (3e^x)^{-4}$

Roots and Rational Exponents

We can expand our use of exponents to include radicals by representing $\sqrt[n]{a}$ as $a^{\frac{1}{n}}$ and $\sqrt[n]{a^m}$ as $a^{\frac{m}{n}}$, where $n > 0$ and m is a real number. We can then apply the rules listed above.

Example 94. Simplify.

1. $\sqrt[3]{27x^6y^9}$
2. $(x^2\sqrt{4x})^2$
3. $c(b^3 + 1)^{\frac{1}{2}}$

Problem 95. Simplify.

1. $\sqrt[4]{x^{12}y^8z^4}$
2. $(\sqrt[3]{m^6n^4})(\sqrt[3]{m^4n^3})$
3. $\frac{\sqrt{r^8t^4}}{\sqrt{r^4t^2}}$

Problem 96. Write each expression below as a sum or difference of powers.

1. $\sqrt{x}(x^3 - 2x - 1)$
2. $\frac{\sqrt{x}y - x^2y^2 + x^3}{x}$
3. $\frac{p^3 - 3p + \sqrt{p}}{\sqrt{p}}$

Notes

3.5 Exercises

1. Let $p(x) = x^2$, $q(x) = -5^x$, $r(x) = 5^{x-3}$. Write a formula for each function below, using the exponent rules to simplify the formula and write without negative exponents.

   (a) $a(x) = p(q(x))$

   (b) $b(x) = 2q(x)r(x)$

   (c) $c(x) = (q(x))^{-3}$

   (d) $d(x) = 5\left(\frac{q(x)}{r(x)}\right)^3$

2. Let $f(x) = \sqrt{x^2}$, $g(x) = \sqrt[3]{x^3}$, $h(x) = \sqrt[4]{x}$, and $k(x) = x^4$.

   (a) True or false: For all real values x, $f(x) = g(x)$. Explain your answer.
   (b) True or false: For all real values $x \geq 0$, $f(x) = k(h(x))$. Explain your answer.
   (c) True or false: For all real values $x \geq 0$, $h(k(x)) = k(h(x))$. Explain your answer.
   (d) True or false: For all real values x, $f(x) = h(k(x))$. Explain your answer.

3. Let $m(x) = x^{\frac{1}{6}}$, $n(x) = x^3$, and $p(x) = x^4$.

   (a) Write $m(n(x))$ as a function of x with a single exponent.
   (b) Write $m(n(x)p(x))$ as a function of x with a single exponent.
   (c) For what values of x is $m(p(x)) = x^{\frac{2}{3}}$? Explain.
   (d) For what values of x is $\sqrt[3]{n(x)} = x$? Explain.

4. (a) For what values of x is $3 \cdot 2^x = 6^x$? Explain.
   (b) Is it ok to simplify $3 \cdot 2^x$ into 6^x? Explain.

5. Write each expression as a sum of powers.

   (a) $\frac{x^3 - 4x^2 + 7x}{\sqrt{x}}$
   (b) $x^{\frac{2}{3}}\left(3x^5 - 6x^{\frac{5}{3}} - 9x\sqrt[3]{x}\right)$
   (c) $x^{\frac{5}{2}}\left(-8x^2 + 5x^{\frac{3}{2}} - 10\sqrt{x} + 4\right)$

6. Let $g(x) = -\frac{1}{2}x^2 + 4x - 6$.

   (a) Find the $x-$ and $y-$intercepts of $g(x)$.

   (b) Find the vertex of $g(x)$.

   (c) Use parts 4a and 4b to graph $g(x)$. Use Desmos to check your answer.

   (d) Solve algebraically: $g(x) = -2.5$. Use Desmos to check your answer.

   (e) Solve algebraically: $g(x) = 1.5$. Use Desmos to check your answer.

   (f) Solve $g(x) < -6$.

7. The graph of $b(n)$ is shown below.

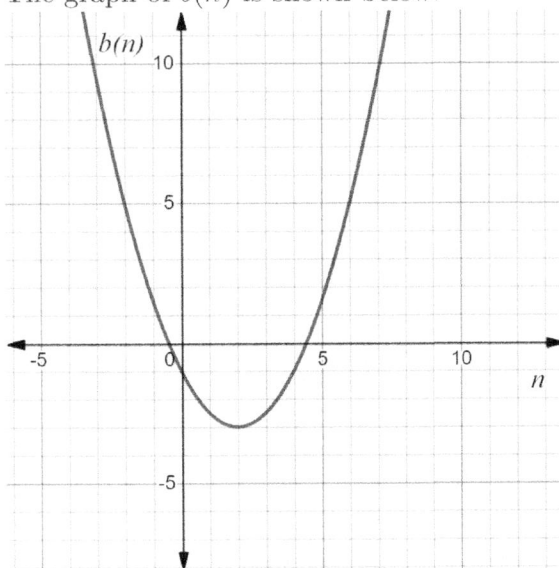

(a) Evaluate $b(0)$.

(b) Solve $b(n) = 5$.

(c) Solve $b(n) \leq 5$.

(d) Write a function equation for $b(n)$.

(e) Use your equation to find the $n-$intercepts of b.

8. The number of computer science degrees awarded by Monroe College has increased by a factor of 1.5 every 5 years since 1984. The college granted 8 degrees in 1984.

(a) Write a formula for the number of degrees awarded t years after 1984.

(b) If the number of degrees awarded continued to increase exponentially, how many degrees were awarded in 2000?

(c) If the number of degrees awarded continues to increase exponentially, in what year will 200 degrees be awarded?

(d) What is the average number of degrees awarded per year during the period from 1984 to 2004?

9. Simplify each expression below and write your answer without any negative exponents.

(a) $\dfrac{7y^6}{4y^5 z^4}$

(b) $(-2a^{-2}b^4)^3$

(c) $\left(\dfrac{2y^4}{4y^2}\right)^5$

(d) $\dfrac{(a^{-1}b^3)^2}{(a^2 b^{-3})^3}$

(e) $\left(\dfrac{wx}{3x^{-3}}\right)^{-2}$

(f) $(xw)(6x^{-6}w^{-4})^3$

(g) $(k \cdot 4k^{-2} \cdot (3k)^2)^3$

10. Simplify each expression below and write your answer without any negative exponents. Assume all variables denote positive numbers.

(a) $x^{\frac{2}{3}} \cdot x^{\frac{4}{3}}$

(b) $(9x)^{\frac{1}{2}} \cdot 4x^{\frac{1}{4}}$

(c) $(27z^3)^{-\frac{2}{3}}$

(d) $(-8x^6y^{-18})^{-\frac{1}{3}}$

(e) $\left(\dfrac{a^{\frac{2}{3}}}{b^{\frac{1}{2}}} \right)^4$

(f) $\sqrt{8p^2q^3r}$

(g) $-2\sqrt[3]{27x^6y^5z^2}$

3.6 Functions and Inverse Functions

Inverse functions are functions that undo the action of a given function.
Goals:

- F: Be able to determine the inverse of a function given given in any form (graph, table, equation).

- F: Be able to determine the domain and/or range of a function given as an equation or a graph.

Definition 97. A **relation** is a set of ordered pairs of the form (x, y). A **function** is a relation where each x value has exactly one output value y.

Example 98. Decide if each relation below, indicated by a graph or by a table, represents a function of x.

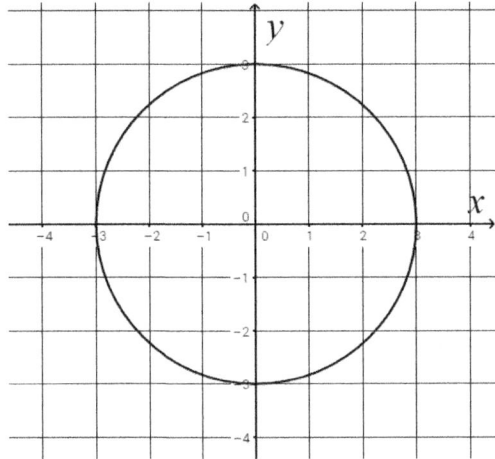

x	y
1	2
3	2
5	2
7	2
9	2

x	y
7	1
3	0
7	-1
5	-2
2	-3

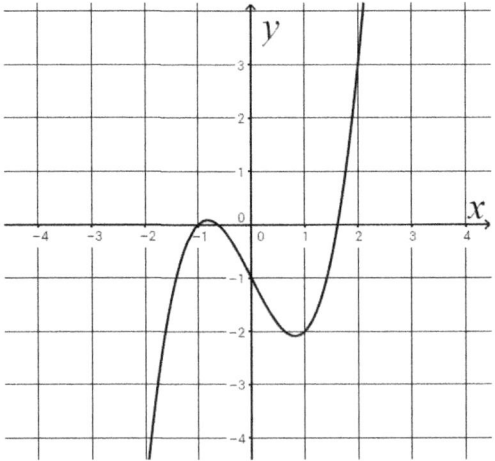

x	$f(x)$
-2	-4
-1	-3
0	-2
1	-1
2	0

x	y
-3	0
-2	-1
-1	2
0	-1
1	3

Definition 99. Given a function f defined on a domain D, a function g on a domain E is an **inverse** of f if $f(g(x)) = x$ whenever x is in the domain of g, and $g(f(x)) = x$ whenever x is in the domain of f. Often, the inverse function g is written f^{-1}.

Example 100. On the domain of all real numbers, $g(x) = \sqrt[3]{x}$ is the inverse of $f(x) = x^3$.

Example 101. For the function $r(x)$ as defined in Table 3.11, the inverse function $r^{-1}(x)$ is the function given in Table 3.12.

x	0	1	2	3
$r(x)$	25	60	1	-10

Table 3.11: $r(x)$

x	25	60	1	-10
$r^{-1}(x)$	0	1	2	3

Table 3.12: $r^{-1}(x)$

Problem 102. Refer to the functions $r(x)$ and $r^{-1}(x)$ in Example 101.

1. What do you notice about the two rows in the tables for the functions?

2. Compute $r^{-1}(r(1))$, $r(r^{-1}(1))$, $r^{-1}(r(2))$ and $r(r^{-1}(25))$.

3. What are the domain and range of r?

4. What are the domain and range of r^{-1}?

Example 103. Consider the function $f(x) = x^2$.

1. Complete the following table of output values of f:

x	-2	-1	0	1	2
$f(x)$					

Table 3.13: f

2. Create a table for a relation, g, that reverses the input and output of f. Is g a function?

3. Since g is not a function, it cannot be the inverse of f. In order for a function to have an inverse, it must be **one-to-one**, that is every output has exactly one input.

4. Sometimes we can restrict the domain of a function to make it one-to-one, so that the restricted function will have an inverse. On what domain would $f(x) = x^2$ have an inverse? What is the range of f on this domain?

5. What is the inverse of f on the domain $[0, \infty)$? What are the domain and range of f^{-1}?

Problem 104. Refer to the function $z(t) = \sqrt{t - 3}$.

1. What are the domain and range of z?

2. Find a formula for the inverse function, z^{-1}.

3. What are the domain and range of z^{-1}?

Problem 105. Refer to the function $h(x) = 3x - 7$.

1. What are the domain and range of h?

2. Find a formula for the inverse function, h^{-1}.

3. What are the domain and range of h^{-1}?

Problem 106. Refer to the functions z and z^{-1} in Problem 104, and h and h^{-1} in Problem 105.

1. For *each* pair of functions, use Desmos to graph them on the same set of axes, along with the line $y = x$. Sketch each graph.

2. What do you notice about the relationship between the graph of a function and the graph of its inverse? Explain why this happens.

Problem 107. What do you know about the relationship between the domain and range of a function and the domain and range of its inverse?

Notes

3.6 Exercises

1. Refer to the function $g(x) = 7x + 2$.

    (a) What are the domain and range of g?
    (b) Let $h(x) = x^2 + x + 1$. Find an algebraic expression for $g(h(x))$.
    (c) Again using $h(x) = x^2 + x + 1$, find an algebraic expression for $3g(x) - h(x)$.
    (d) Find a function equation for the inverse function, g^{-1}.
    (e) What are the domain and range of g^{-1}?

2. Refer to the graph of $R(x)$ in Figure 3.2. The diagonal line $y = x$ is included on the graph for reference. Note that the graph of $R(x)$ includes the points $(-7.5, 0)$, $(-1, 2)$, $(0, 4.5)$, and $(6, 6)$.

    (a) What are the domain and range of R?

    (b) Draw the graph of the inverse function, R^{-1}.

    (c) What are the domain and range of R^{-1}?

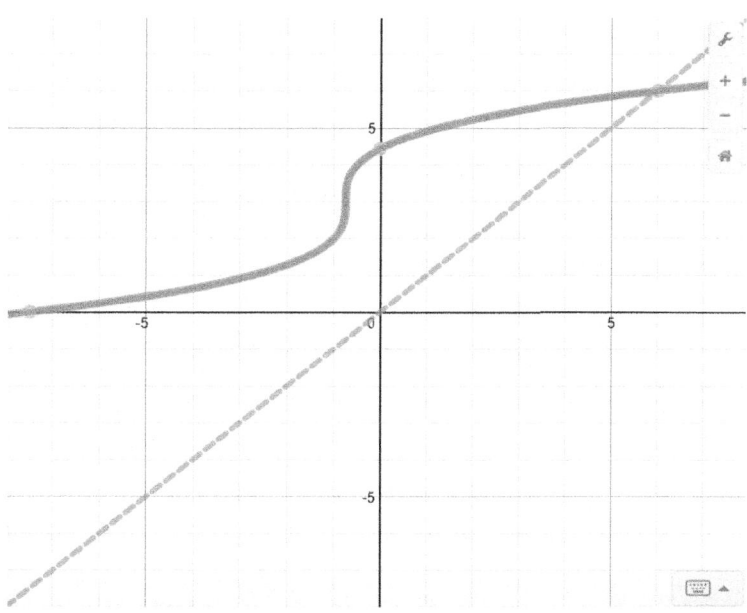

Figure 3.2: Graph of $R(x)$

x	0	1	2	3
$f(x)$	3	0	2	-4

Table 3.14: $f(x)$

3. Refer to f in Table 3.14.

    (a) What are the domain and range of f?
    (b) Find a function table for the inverse function, f^{-1}.
    (c) What are the domain and range of f^{-1}?

4. Let $A(x) = 4(x-6)^{\frac{3}{2}}$.

    (a) Write functions $b(x)$ and $c(x)$ so that $A(x) = b(c(x))$.
    (b) What are the domain and range of $A(x)$?
    (c) Find a function equation for the inverse function, $A^{-1}(x)$.
    (d) What are the domain and range of $A^{-1}(x)$?
    (e) On what interval(s) is A increasing? On what interval(s) is A decreasing?

5. Let $Z(x) = (x^2 + 4)^{\frac{1}{2}}$, $x \geq 0$.

    (a) True or false: For all $x \geq 0$, $Z(x) = x + 2$. Explain your answer.
    (b) What are the domain and range of $Z(x)$?
    (c) Find a function equation for the inverse function, $Z^{-1}(x)$.
    (d) What are the domain and range of $Z^{-1}(x)$?

6. Refer to $Q(x) = \sqrt[3]{x - 3}$.

    (a) What are the domain and range of Q?
    (b) Let $U(x) = x^3$. Find an algebraic expression for $U(Q(x))$.
    (c) Find a function equation for the inverse function, Q^{-1}.
    (d) What are the domain and range of Q^{-1}?

7. The Jaguar XJ AWD loses value more rapidly than some of its competitors. From the initial purchase price of \$76,700, an owner can expect the value of the car 5 years later to be \$52,014.

    (a) What is the average dollar value decline per year during the first 5 years of ownership?
    (b) Use an exponential model produce a function that gives the value of the Jaguar in terms of the number of years, t, since the car was new.
    (c) Use your model to predict the value of the Jaguar at 3 years old.
    (d) When will the value of the Jaguar decline to \$30,000?

8. Simplify each expression below and write without negative exponents.

    (a) $\frac{xy^7}{x^3 y^4}$

    (b) $\left(\frac{2X^3}{-8X^4} \right)^2$

    (c) $\frac{8x^3(x^2)^8}{(-2x^6 y)^2}$

3.7 Logarithms

Goals:

- E: Be able to solve an equation with an unknown exponent.

- E: Be able to model a situation with appropriate exponential equation(s) and interpret the solution.

- E: Be able to use definition and properties of logarithms to rewrite expressions involving logarithms in different forms.

- F: Be able to determine the inverse of a function given given in any form (graph, table, equation).

Logarithms are used to solve for unknown exponents in equations.

Definition 108. If x is a positive number then $\log_b(x)$ is the exponent of b that gives x. That is

$$y = \log_b(x) \qquad \text{if and only if} \qquad b^y = x$$

The number b is called the base of the logarithm and is always greater than 0.

Definition 109. The **natural logarithm**, written as $\ln x$, is the logarithm with base e.

Investigation 110. The number e.

1. Use a calculator or Desmos to evaluate the following:

   (a) $1 + \frac{1}{1!}$

   (b) $1 + \frac{1}{1!} + \frac{1}{2!}$

   (c) $1 + \frac{1}{1!} + \frac{1}{2!} + \frac{1}{3!}$

   (d) $1 + \frac{1}{1!} + \frac{1}{2!} + \frac{1}{3!} + \frac{1}{4!}$

   (e) $1 + \frac{1}{1!} + \frac{1}{2!} + \frac{1}{3!} + \frac{1}{4!} + \frac{1}{5!}$

   (f) $1 + \frac{1}{1!} + \frac{1}{2!} + \frac{1}{3!} + \frac{1}{4!} + \frac{1}{5!} + \frac{1}{6!}$

2. Use a calculator or Desmos to evaluate the following:

   (a) $\left(1 + \frac{1}{1}\right)^1$

   (b) $\left(1 + \frac{1}{2}\right)^2$

   (c) $\left(1 + \frac{1}{3}\right)^3$

   (d) $\left(1 + \frac{1}{4}\right)^4$

   (e) $\left(1 + \frac{1}{100}\right)^{100}$

   (f) $\left(1 + \frac{1}{1000}\right)^{1000}$

3. Find the value of e on your calculator or using Desmos. How does this compare to what you found above?

Problem 111. Compute the following *without* using a calculator.

1. $\ln(e^2)$

2. $\ln(e^6)$

3. $\log(10^3)$

4. $\log_2(2^{-2})$

5. $\ln(e)$

6. $\ln(1)$

7. $\ln(e^a)$

Investigation 112. We will explore the relationship between log and exponential functions. A graph and table for $g(x) = \log_2(x)$ are given below.

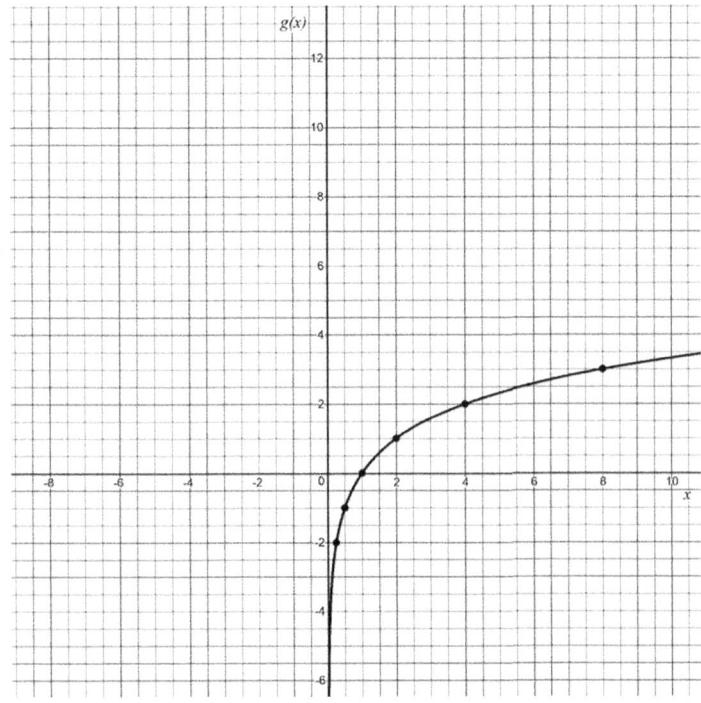

x	$\log_2 x$
.25	-2
.5	-1
1	0
2	1
4	2
8	3

1. Use the graph above to construct the graph of the inverse function, $g^{-1}(x)$.

2. Use the table above to construct a table of the inverse function, $g^{-1}(x)$.

3. Based on your graph in part 1 and your table in 2 above, write an equation for the inverse function, $g^{-1}(x)$.

4. What are the domain and range of $g(x)$?

5. What are the domain and range of $g^{-1}(x)$?

6. Explain why $g(x) = \log_2(x)$ is not defined for $x \leq 0$.

For any base, the inverse of the exponential function $f(x) = b^x$ is $f^{-1}(x) = \log_b(x)$, so that for positive x values, $b^{\log_b(x)} = x$, and for all real numbers, $\log_b(b^x) = x$. For example, the natural logarithm, $ln(x)$, is defined on the domain of positive real numbers $(x > 0)$, and e^x is defined on the domain of all real numbers. Since $\ln(e^x) = x$ and $e^{\ln x} = x$, these functions are inverses of each other.

Problem 113. Use the definition of logarithm to solve the following equations.
1. $145 = e^{2z}$ 2. $0 = 84 - 3 \cdot 10^{-x}$

3. $-4 + 3\log_3 x = 5$

4. $3\ln(x - 1) = 8$

Problem 114. The population of bacteria in a lab culture doubles every 15 minutes.

1. If there are initially 1000 bacteria in the culture, write a function, $P(t)$, that gives the number of bacteria in the culture at time t minutes.

2. How many bacteria will be present after 1 hour?

3. How long will it take the bacteria to reach 20,000?

4. Find an equation for the inverse function that gives the time, t, as a function of the population, P.

Notes

3.7 Exercises

1. Solve each equation below.

   (a) $k(x) = 3e^{2x} + 1$, solve $k(x) = 11$ for x

   (b) $q(x) = -6\log_3(x - 3)$, solve $q(x) = -24$ for x

   (c) $f(x) = 2\log_6 4x$, solve $f(x) = 4$ for x

   (d) $g(x) = 7^{2x-3} - 4$, solve $g(x) = 14$ for x

   (e) $a(x) = 5^{3-2x}$, $b(x) = 5^{-x}$, solve $a(x) = b(x)$ for x

   (f) $c(x) = 8^{x-1}$, $d(x) = 2^{x+2}$, solve $c(x) = d(x)$ for x

2. Solve for t in the equation $A = 25 \cdot 3^{t/8}$.

t	0	10	20	30
$P(t)$	331,767	386,988	450,557	545,852

Table 3.15: Population, $P(t)$, of Albuquerque t years after 1980

3. Refer to the population of Albuquerque, NM, for given years as shown in Table 3.15.

   (a) Using the population of Albuquerque in 1980 and 2010, build an exponential model for $P(t)$.

   (b) Use your exponential model to predict the population of Albuquerque in 1990. How does your prediction compare with the actual population at that time?

   (c) Use your exponential model to predict in what year the population of Albuquerque will be 700,000. First give your answer in exact form using a logarithm, and then give the decimal approximation.

   (d) Using the population of Albuquerque in 1980 and 2010, build a linear model for $P(t)$.

   (e) Do you think the linear model or the exponential model is a better fit to the data? Explain.

4. The population of Fresno, CA was 430,724 people in 2000. At that time, the population was growing at an annual rate of 1.3%.

   (a) Write an exponential equation relating the population, $P(t)$, to the number of years, t, with $t = 0$ corresponding to 2000.

   (b) If the population continued to grow at that rate, what is the population of Fresno in 2015?

   (c) According to the model, when will the population of Fresno reach 600,000?

5. Shayla opens a banking account with $300. The account earns 2.2% compounded annually. Let t be the number of years the bank account has been open, and $A(t)$ the balance in the account.

   (a) Make a table showing the balance, $A(t)$, in the account at $t = 0, 1, 2, 3$, and 4 years.

   (b) Write an equation for the function $A(t)$.

   (c) Use your function equation to determine the balance in 20 years.

   (d) When will the balance double?

6. Solve. Give your answer in both exact form and as a decimal approximation.
   (a) $6 \cdot 3^{8x} = 1000$ 　　　　　　　　　　　　(b) $10^{-2v} + 365 = 2118$

7. The intensity of light d meters away from a 100-Watt bulb, measured in Watts per square meter (W/m^2), can be modeled by the equation

$$I = a \cdot d^{-2}$$

   (a) If the intensity of light 1.5 meters from the bulb is 3.53 W/m^2, find the constant a and write the equation to model the intensity of light d meters away from a 100-Watt bulb.

   (b) What will the intensity of light be 2.5 meters away from the bulb?

   (c) At what distance away from the bulb will the light have an intensity of 4 W/m^2.

   (d) Assuming that both distance and light intensity are positive, write an equation for the inverse function that gives the distance from a 100-Watt bulb as a function of the light intensity measured at that distance.

8. Simplify each expression.

   (a) $2\sqrt[3]{x^2}(x^4 - 5x^3 - x)$

   (b) $\dfrac{x^{-2} - x + 3x^2}{x^{-2}}$

3.8 Properties of Logarithms

Goals:

- F: Be able to solve an equation with an unknown exponent.

- F: Be able to use definition and properties of logarithms to rewrite expressions involving logarithms in different forms.

Investigation 115.

1. We will use the exponent rule for products, $a^p \cdot a^q = a^{p+q}$, to prove that $\log_a(xy) = \log_a x + \log_a y$ where a, x and y are greater than 0 and $a \neq 1$.

2. We will use the quotient rule of exponents, $\frac{a^p}{a^q} = a^{p-q}$, to prove that $\log_a\left(\frac{x}{y}\right) = \log_a x - \log_a y$ where a, x and y are greater than 0 and $a \neq 1$.

3. We will use the power rule of exponents, $(a^p)^q = a^{pq}$, to prove that $\log_a x^k = k \log_a x$ where a and x are greater than 0, $a \neq 1$ and q is a real number.

Problem 116. Rewrite each expression using a single natural logarithm.

1. $2 \ln u + \frac{1}{3} \ln v - 3 \ln w$
2. $\frac{1}{2}[\ln x - 2 \ln(x+1) - \ln(2x+3)]$

Problem 117. Use the properties of logarithms to expand the expression as a sum or difference of logarithms.

1. $\log_2(2x^3)$
2. $\log \frac{5}{\sqrt{x}}$
3. $\ln \frac{x^4 \sqrt{z}}{y^3}$

Problem 118. Solve $5^{2x} = 28$ two ways below.

1. Solve using the definition of logarithm.   2. Solve using inverse functions and properties of logarithms.

3. How are the two solutions above the same? How are they different?

Problem 119. Solve each equation using inverse functions and properties of logarithms.

1. $3e^{\frac{1}{2}x} + 7 = 16$ 2. $4 \cdot 10^{x-5} = 460$

3. $\ln(5x - 4) = 2$ 4. $\log_4 x - \log_4(x - 1) = 2$

Notes

3.8 Exercises

1. Rewrite the expressions using a single natural logarithm.
   (a) $\ln a + 4\ln b - 3\ln c$
   (b) $\frac{1}{2}\ln s - 2\ln t - \frac{1}{2}\ln c$

2. Solve for x in the following equation: $\ln(x-2) + \ln(x-5) = \ln 4$.

3. Solve for t in the following equation: $3^{t-1}7^{t-2} = 27783$.

4. Solve $e^{t^2-2t-10} = 200$.

5. Solve. Give your answer in exact form.
   (a) $2\ln(x-2) - \ln x = 1$
   (b) $\log_{10}(x-3) + \log_{10}(x+5) = 1 + \log_{10} 2$
   (c) $e^{2x} - 3e^x = 10$
   (d) $7^{2t} + 4\cdot 7^t - 32 = 0$

6. Recall that any object experiencing the force of gravity can be modeled by the equation $h(t) = -16t^2 + vt + c$, where t is the time in seconds, $h(t)$ is the height in feet, c is the initial height of the object, and v is the initial velocity of the object. A soccer ball is kicked from a height of 2 feet above the ground and has an initial velocity of 70 feet per second.

   (a) What is the highest the ball will go?
   (b) During what time interval is the ball at least 10 feet in the air?
   (c) How long is the ball in the air?

7. Let $f(x) = 2e^x$, $g(x) = 2x - 1$ and $k(x) = x^2$.

   (a) Evaluate $f(g(-1))$.
   (b) Write an expression for $(k \circ g)(x)$.
   (c) Write an expression for $g(f(x)) - k(x)$.
   (d) Write an expression for $f(g(k(x)))$.
   (e) For what values of x does $g(x) = k(x)$?
   (f) For what values of x is $f(x) \le k(x)$?

8. Refer to the function $g(x) = -\frac{1}{3}x + 4$.

   (a) Find a function equation for the inverse function, g^{-1}.
   (b) What are the domain and range of g^{-1}?
   (c) Graph $g(x)$ and $g^{-1}(x)$ on the same set of axes.
   (d) Solve $g(x) > 5$.

Made in the USA
Columbia, SC
12 September 2025

61928017R10076